ETHICAL ISSUES IN GOVERNI

Ethical Issues in Governing Biobanks

Global Perspectives

Edited by

BERNICE ELGER
University of Geneva, Switzerland

NIKOLA BILLER-ANDORNO
University of Zurich, Switzerland

ALEXANDRE MAURON
University of Geneva, Switzerland

ALEXANDER M. CAPRON
University of Southern California, USA

Routledge
Taylor & Francis Group

LONDON AND NEW YORK

First published 2008 by Ashgate Publishing

2 Park Square, Milton Park, Abingdon, Oxon OX14 4RN
711 Third Avenue, New York, NY 10017, USA

Routledge is an imprint of the Taylor & Francis Group, an informa business

First issued in paperback 2016

British Library Cataloguing in Publication Data
Ethical issues in governing biobanks : global perspectives
    1. Human genetics - Databases - Moral and ethical aspects
    2. Human genetics - Databases - Law and legislation
    I. Elger, Bernice
    174.2'8

Library of Congress Cataloging-in-Publication Data
Ethical issues in governing biobanks : global perspectives / edited by Bernice Elger ... [et al.].
    p. cm.
    Includes bibliographical references and index.
    ISBN-13: 978-0-7546-7255-5 (hardback)
    ISBN-10: 0-7546-7255-7 (hardback) 1. Human genetics--Databases--Moral and ethical aspects. 2. Human genetics--Databases--Law and legislation. I. Elger, Bernice.
    [DNLM: 1. Genome, Human. 2. Confidentiality. 3. Databases, Genetic--ethics. 4. Databases, Genetic--legislation & jurisprudence. 5. Genetic Research--ethics. 6. Genetics, Population--ethics. QU 470 E835 2008]

    QH438.7.E836 2008
    174.2--dc22

                                                            2008003427

ISBN 13: 978-0-7546-7255-5 (hbk)
ISBN 13: 978-1-138-26244-7 (pbk)

Contents

Notes on the Contributors

Ma'n H. Abdul-Rahman is Research Assistant at the Centre de Recherche en Droit Public (CRDP), Université de Montréal, Canada.

Andrea Boggio is Assistant Professor of Legal Studies, Bryant University, USA.

Nikola Biller-Andorno is Professor of Biomedical Ethics and Director of the Institute of Biomedical Ethics, University of Zurich, Switzerland.

Alexander Morgan Capron is University Professor, Scott H. Bice Chair of Healthcare Law, Policy and Ethics, and Co-Director, Pacific Center for Health Policy and Ethics, University of Southern California, USA.

Bernice Simone Elger is Professor at the Institute of Legal Medicine (unit for health law and medical ethics), University of Geneva, Switzerland.

Agomoni Ganguli-Mitra is Research Assistant at the Institute of Biomedical Ethics, University of Zurich, Switzerland.

Bartha Maria Knoppers is Professor of Law, Faculté de Droit, and Canada Research Chair in Law and Medicine, Centre de Recherche en Droit Public (CRDP), Université de Montréal, Canada.

Alexandre Mauron is chair of the Institute for Biomedical Ethics, University of Geneva, Switzerland.

Effy Vayena is Senior Research Fellow at the Institute of Biomedical Ethics, University Priority Program in Ethics, University of Zurich, Switzerland.

Acknowledgements

We gratefully acknowledge the financial support of the Geneva International Academic Network and a publication subsidy from the Faculty of Medicine of the University of Geneva.

We would like to express our gratitude to all the interviewees who shared with us their views about the ethical and regulatory aspects of genetic databases in a global context.

Chapter 1

Introduction:
Biobanks, Genomics, and Research—
A Nightmare for Public Policy Makers?

Alex Mauron

1. From the Human Genome to human genomes

In the early nineteen nineties, impressed by the increasing speed and automation of DNA sequencing, as well as progress in an emerging field that was not yet called bioinformatics, a few visionaries saw the complete deciphering of the human genome as a realistic goal within a not-too-distant future (Sulston and Ferry 2002). Finding the "Holy Grail" of biology was anticipated as a crowning and spectacular achievement, marking a new era of biological understanding for mankind (Gilbert 1992). It is hard nowadays to capture the sense of awe before the task at hand and the bitter controversies about its feasibility that prevailed in those days. This is because by the time the complete sequence of the human genome was available in 2003 (Collins, Green et al. 2003), scientists were already taking the human genome for granted. Its sequence became integrated as background knowledge for further research, that soon moved on to other "omes": proteomes, transcriptomes, metabolomes … (Petsko 2002). In addition, human genomics gave a new lease on life to human population genetics. The technologies that had been developed to read the generic "Book of Man" were soon redirected towards specific genetic databases, or biobanks,[1] representing collections of individual genomes. The purpose became to map precisely the differences between genomes, in particular to assess the most common form of variation, namely single nucleotide polymorphisms (HapMap 2005). That is the current stage in the development of genomics with which this book is mainly concerned as regards its ethical and legal implications.

Large scale longitudinal studies as well as collections of biological materials from more or less extensive populations were established and new projects along these lines are continuously initiated today (Cambon-Thomsen 2003). One of their goals

---

1 In this volume, the terms "biobank" and "genetic database" are used interchangeably to signify a collection of human biological samples that can be used for genetic analysis, including those that combine such samples with the results of genetic analyses and health or other data about the persons from whom the samples were collected. The category encompasses pathology collections, repositories for specific diseases (e.g. cancer registries), and population databases created to permit longitudinal studies of any disease or condition.

is to connect data on genetic variation with differential susceptibility or resistance to many diseases. The hope is for a more individualized preventive medicine, but also for new knowledge about pathogenic mechanisms as well as the identification of therapeutic targets for innovative drugs, as the pharmaceutical industry was quick to realize. Today, a large number of biobank research projects are under way, which differ in scale, scientific objectives, methodology, and most importantly perhaps, in their public or private nature.

Early on, ethical commentary and public policy discussions emerged, as it was soon clear that the new biobanks presented a challenge to the classical ethical frameworks for research on human subjects and for medical genetics respectively (Chadwick and Berg 2001). For instance, research ethics entails the right of participants to be fully informed of the objectives and procedures of a well specified research project, and the right to withdraw from that project at any time. This sits uneasily with a biobank's typically open-ended scientific goals and the logistic difficulties—or sheer impossibility in many cases—of recontacting the people whose biological material and related data are stored in biobanks every time a new research project intends to make use of their samples (Lipworth, Ankeny et al. 2006). Biobanks also did not align with the ethical norms of medical genetics which recognize the familial nature of genetic information yet still emphasize the individualistic, autonomy-based perspective that is inherent in non-directive counselling, and the expectation of individual benefit or protection from personal harm. This differs from the populational logic of biobanks and raises the question of how (if at all) incidental findings relevant to an individual participant's health may usefully be conveyed to that person. Such harm-preventing feedback to participants may seem mandated by the principle of non-maleficence yet would contradict the ethical imperative of privacy protection as embodied in the requirement to anonymize data, which is one reason why that requirement is increasingly criticized (Kohane, Mandl et al. 2007).

Much of the ethical and policy issues raised by biobanks revolve around the tension between individual rights and claims enshrined in classical bioethical principles regarding human subject research and human genetics, and the populational outlook and objectives of biobank research. This tension is manifest in all areas of ethical concern, such as informed consent, confidentiality and privacy, ownership of samples and management of commercial interests. As a result, one of the major conceptual questions raised by biobank research is how, and to what extent, research ethics needs to be supplemented or revised by taking on board public interests that go beyond the individualistic focus of autonomy-based principles (Knoppers 2005). How this extension and/or revision should go about is a recurring feature of the research reported in this book, as well as of the ethical and policy commentaries in following chapters.

2. Linking genomes with health data

Several research strategies are involved in human genetic biobanks (Mc Fleming 2007). Linkage studies search for gene variants linked with disease susceptibility in particular families, whereas association studies address common diseases in large populations.

Whatever their differences in scale and methodology, most of these studies link data about genetic variation between individuals and groups with information on specific diseases, disease susceptibility, drug responsiveness, or other health-relevant traits, including environmental and behavioural variables. This highlights the importance of confidentiality for genetic data as well as for medical and lifestyle information. As to both, unauthorized disclosure infringes the right to privacy and exposes a person to possible harm such as discrimination and stigmatization (Anderlik and Rothstein 2001; McGuire and Gibbs 2006). Biobank research poses the additional potential problem that the harm resulting from a breach of confidentiality may be difficult to anticipate if it stems from some novel insight into the association between genes and health data that is produced by the research itself. The requirements of confidentiality and privacy aim to protect research participants' interest in exerting some control over the consequences of their participation, especially to avoid research findings and data backfiring against them. It is therefore not surprising that many discussions in the field, as well as many regulations, address the technical and ethical validity of measures, such as coding and anonymization, designed to protect confidentiality and privacy (see Chapters 2 and 3).

The notion of a controlling interest of biobank research participants also raises the question of what ownership interest (if any) they have in their samples and associated data. As discussed further in the next chapter, a straight property approach is largely agreed to be impractical and misguided. Nevertheless, the broader issues behind the concept of ownership remain and may even require revising the concept of property (Bjorkman and Hansson 2006). What is the appropriate relationship between the samples/data with (1) the individual providing a sample? (2) the biobank? (3) the researchers using the biobank? (4) other possible stakeholders, such as the community to which the research participants belong? What sort of controlling interest do participants have over their samples and data? Through what kind of societal arrangement is the public interest vested in biobank research supposed to materialize? What concept of benefit-sharing is the most appropriate? What intellectual property interests are legitimate? These questions soon move the debate away from traditional bioethical discussions to issues of social ethics and economic policy.

3. The genomic Tower of Babel

The Tower of Babel metaphor has been invoked for the confusing terminology used in privacy and confidentiality discussions to describe how the link between an individual (a "sample source") and a given biobank sample and associated data will be obscured (Knoppers and Saginur 2005; Elger and Caplan 2006). Samples and data are described by a bewildering array of terms: *identified* (no obscuring at all), *coded* (link kept secret to various extents), *anonymized* (link irretrievably severed), and *anonymous* (no link in the first place, or so it seems), with a huge assortment of confusing and sometimes contradictory quasi-synonyms for each of these terms. In part, this reflects the large number of organizations which, over time, took it upon themselves to review this area and propose regulations. However, if that were the whole problem, a concerted effort towards semantic clarification and the elimination of unwanted synonymy would clear up the babble.

Yet there are other sources of complexity, if not confusion. One is that the degree of protection afforded by various coding techniques must be assessed against existing or potential technical possibilities to break the code and identify study participants (Lin, Owen et al. 2004). This leads to uncertainties about levels of data security that are both achievable and desirable. Moreover, the terminological confusion probably points to a more basic uncertainty about what level of secrecy best protects the personal interests expressed in the concepts of confidentiality and privacy, while allowing the research to proceed without excessive impediments. This, rather than the technicalities of coding and anonymizing, should be the focus of ethical analysis. Indeed, one of the frustrating aspects of these debates is the difficulty of distinguishing genuine ethical differences from mere disagreements about feasibility.

This complex terminology is also linked to a common but poorly articulated perception that genetic data are in some way different, and more sensitive, than "ordinary" medical data. As pointed out by Knoppers and Saginur, this genetic exceptionalism is rarely defended explicitly but rather fueled unwittingly by regulatory efforts that focus exclusively on genetic data without adequately considering their compatibility with the management of medical data in general (Knoppers and Saginur 2005). On the other hand, one has to recognize that genes and DNA are indeed special in some sense, though that does not justify exceptional treatment of the medical information that originates with them. After all, DNA is "stuff." DNA samples, or biological samples containing DNA, are fully material realities, and as such they invoke the language of material possession. This is why proprietary interests in biobank samples need to be clarified. Yet DNA is information as well, either explicit (as in the genotyping data obtained from it in a given research project) or implicit (since a DNA sample can be "interrogated" to reveal more, or all, of its informational content). The informational aspect calls for the language of data sharing, in furtherance of the public interest in knowledge, but also the language of data withholding, with a view either to protecting privacy or to securing the intellectual property interests of the person who has produced the information. Information travels light: It can be exchanged without loss, unlike the material substance whence it originates. Information can either be protected by intellectual property law, in which case it may be sold and bought just like material possessions, or it can be treated as a public good, which may increase its utility considerably, again unlike most material possessions. Past discussions about the knowledge generated by the Human Genome Project are illuminating in this respect. At one point, two models were competing (Sulston and Ferry 2002). One proposed that genome data obtained by a private sequencing consortium should be made a marketable commodity and that access rights to the Human Genome database should be sold to researchers and companies. The other view, enshrined in the Bermuda agreement, posited the human genome sequence to be a public good and demanded that validated genome data should be publicly accessible without restriction (Marshall 2001). The second view prevailed, and since the Human Genome is now as basic to human genetics as Mendeleyev's table of chemical elements is to chemistry, it is hard to imagine how it could have been otherwise.

4. The fog lifts

"An ethical patchwork" is how one ethicist described the regulatory scenery regarding biobanks several years ago (Maschke 2005), and this confusing situation still prevails. Just as too many cooks are said to spoil the broth, the multiplicity of institutional actors poised to analyze and regulate biobanks explains much of this unfortunate situation. In addition to the semantic thicket mentioned earlier, the diversity of biobanks also contributes to the difficulty of mapping the field and its problems. For instance, there is a distinction to be made—with some overlap—between the issues raised by newly established biobanks for research (which is the main but not sole topic of this book[2]), and collections of biological samples that have originated from routine clinical practice (for example, the specimens collected by a pathology department) and whose research potential as biobanks appears as an afterthought. Clearly the issue of informed consent is quite different in the latter case, since no consent to research was obtained initially to establish these collections. Unsurprisingly, there are "meta-debates" about what it is exactly that biobank regulations should cover and what the important ethical and policy questions really are (Cambon-Thomsen, Rial-Sebbag et al. 2007). This is also frustrating, and it is not unusual for anyone trying to make sense of the literature and the controversies to have difficulty in seeing exactly what different discussants are actually disagreeing about. In addition, there is a huge variety of laws and regulations that have some relevance for biobank research: human subject research regulations, data protection legislation, health insurance legislation, medical law, general principles of civil law, and so forth. As a result, the field has become a subject for complex legal scholarship, and is accordingly difficult to generalize across national boundaries (McGuire and Gibbs 2006).

Nevertheless, the field has moved a long way from the highly controversial debates of the late nineties, with their sweeping and sometimes unfocused criticisms of biobank research and confusing attempts to "reinvent the wheel" of ethical reasoning to cope with this new biomedical research tool. In fact, both in the reports of research and in the commentaries that follow, the reader will discern genuine progress and areas where consensus is emerging. One such area concerns coding, now usually deemed preferable to anonymization. In fact the coding vs. anonymizing debate may eventually be made moot by technological advances. More sophisticated data processing strategies may produce the best of both worlds: robust privacy protection shielding research participants from unwanted disclosure and stigmatization; but also flexibility in opening channels of communication to particular participants that could benefit from specific incidental findings, without compromising data protection for the whole cohort (Kohane, Mandl et al. 2007). In the future, the still common and expedient policy of "no individual feedback" may eventually lose its initial appeal, just as anonymization is losing it now.

Another area of possible consensus is emerging, namely in favor of relatively simple and broad consent procedures, balanced by ethics oversight of new research projects and a strong duty of biobanks to maintain communication and provide

---

2 But see scenario D, Chapter 4.3.

timely and updated information to the study population. Last but not least, it is now much clearer that good governance of biobank projects, especially large scale ones, entails extensive public consultation and oversight by genuinely independent bodies. The efforts and resources needed to ensure enduring public trust cannot be overestimated (Tutton, Kaye et al. 2004). New initiatives, such as the P3G consortium (Public Population Projects in Genomics) are expected to raise the general standards of biobank research, facilitate the integration of ethical issues and foster public participation (Cambon-Thomsen, Rial-Sebbag et al. 2007). They are part of a general trend towards harmonization of large scale projects, also as regards ethical and regulatory issues.

5. The future: from human genomes to your genome and mine

The ultimate step in genomics would, of course, be the complete sequencing of the genome of named individuals, an achievement that is now reaching practical feasibility. Several groups have announced projects to that end, including a private company that has launched "Project Jim": the sequencing of the genome of James Watson, co-discoverer of the double helix structure of DNA and Nobel Prize winner (Marshall 2007). It is reported that using the Human Genome as a template, a specific human genome can be sequenced with high accuracy within weeks at a cost of about $1 million. In that particular case, anonymity was not an issue and the illustrious experimental subject agreed that his genome data could be made public. Interestingly, he did request however that the status of his apolipoprotein E genes be blanked out (specific alleles on that locus can predispose to Alzheimer's disease and cardiovascular problems). Once such person-specific genomic analysis is a reality, one could say that the bioethical discussion will have turned full circle. From the romantic, "pregenomic" stage of the eighties and nineties, when scholars were pondering what it would mean for mankind to access the "Book of Life" (Keller 1995), the "postgenomic" debate has become much more focused and precise, because biobanks posed very specific and practical questions. Bioethics tried to follow the increasingly complex ramifications of these questions in terms of ethical principles, professional and institutional regulations, and multiple legal frameworks. This is the rather technical and sometimes confusing state of the debate that this book reflects. But "Project Jim" and other similar endeavours already point to a new situation, in which genomics will obtain truly comprehensive genetic information about designated individuals. This will probably necessitate some redefinition of the issues revolving around privacy, confidentiality and genetic discrimination, if only because the alluring comprehensiveness of "total" genomic information will weaken arguments against disclosure to participants. But on a more profound level, privacy and control over one's genetic information and its personal significance will become (again) a highly personal and existential matter. Aside from the public policy issues that are at the forefront today, ethicists will have to return to reflecting on the more intimate and individual questions confronting human beings as research subjects, but perhaps also as "consumers" in various guises of their own complete genetic information. But that is another story.

6. The research project and this book

This book has its roots in a collaborative research project undertaken jointly by the Department of Ethics, Trade, Human Rights and Health Law of the World Health Organization (Geneva), and the Institute for Biomedical Ethics, Faculty of Medicine, University of Geneva. The project, entitled "Towards a global ethical framework for research biobanks: a qualitative study among international and US experts," was aimed primarily at analyzing the situation of biobank research as regards its ethical and regulatory framework, with a view to contributing to its improvement in the future. This was achieved by taking stock of the existing scholarly literature and regulations, as well as by using a qualitative research methodology to explore the views of individuals professionally involved with biobank research as well as with the related ethical and legal issues. The method chosen reflected the exploratory nature of the project, which tried to collect a wide range of informed views from persons who examine the issues from a variety of perspectives. This was done by semi-structured interviews of two groups of respondents, who were asked to react to four scenarios illustrating several ethical issues in biobank research. (These scenarios were designed by the project group on the basis of our analysis of existing regulations and debates in the literature). One group of interviewees, called thereafter the "international sample," included individuals from many world regions (including North America), whereas the second group was US-based. Our central concern in conducting this qualitative research was to obtain a rich harvest of considered opinions, which would not only reflect different types of professional engagement with the issues, but also a genuine diversity of regional and cultural backgrounds.

The project was funded mainly by a grant of the Geneva International Academic Network (GIAN) to Alex Capron and Alex Mauron, with additional financial support from WHO and the University of Geneva. The grant from GIAN funded a postdoctoral stay for Andrea Bioggio in Geneva (2004-06), who conducted the interviews with the international sample. The interviews with US respondents were conducted by Bernice Elger during a stay at the University of Pennsylvania in 2004-05 supported by a training grant from the Swiss National Science Foundation. The research team also included Nikola Biller-Andorno at WHO (now at the University of Zurich) and Agomoni Ganguli-Mitra, originally a WHO intern and now a graduate student with Prof. Biller-Andorno. A high-level expert meeting, also funded by GIAN, WHO, and the University of Geneva, was convened towards the end of the project to discuss its results and the broader issues of biobank regulation. Specific chapters in this book describe the project in more detail and provide in-depth analyses of the data. They also include additional background material and commentaries from members of the research group as well as invited contributors. Additional information about the project can be found on the GIAN website (GIAN 2007).

In Chapter 2, Knoppers and Abdul-Rahman provide a comprehensive review of the literature on biobanks, organized around three major issues: consent, confidentiality and commercialization. The discussion of consent includes the various options that have been proposed to address the question of secondary use of samples, as well as the reinterpretations needed for the right to withdraw to remain a practical possibility in the context of biobanks. Confidentiality raises the questions of coding and anonymization, which in turn implies a discussion of the possible informational return to study participants. Finally the section on commercialization gives an overview of the various property claims involved in biobank governance as well as the issue of benefit-sharing. Chapter 3 describes the wealth of guidelines addressing various aspects of biobank research and provides an in-depth analysis of current controversies. The authors map out the uneven regulatory density on various issues, pointing out some topics that have had relatively little attention, such as the problem of group involvement in consent procedures. Chapter 4 describes the qualitative research project that is the central feature of this book. The chapter includes a detailed presentation of the methodology and of the characteristics of both the international and the US samples. Chapter 5 presents the results in more detail and includes, for both groups of respondents, (1) a description of responses on specific items, (2) the identification of areas of consensus or disagreement, (3) an analysis of argumentative patterns or positions, and of their consistency, (4) a discussion of explicit or hypothetical reasons for the positions, and (5) conclusions from the findings for guidelines concerning each specific item. Chapter 6 presents a summary of the high-level expert meeting convened in Geneva on May 8-9, 2006, to discuss the research project and its results, as well as to exchange views about biobank regulation, a meeting that also pointed to areas of significant consensus. In the final chapter, Alex Capron discusses the different views of biobank regulation and offers a general assessment of the future of international guidelines in this field.

At the time when this research project was underway, "a policy nightmare" was a realistic characterization of the field of genomics and biobanks. Today the challenges remain formidable, especially considering that the science is not standing still to let ethicists, lawyers and regulators quietly make up their minds. But the overall impression has changed. There seems to be more conceptual clarity about the issues, a few important areas of consensus, and more sophistication in remaining disputes. Not the matter of rosy dreams, perhaps. But we awoke from the nightmare.

Bibliography

Anderlik, M.R. and Rothstein, M.A. (2001), "Privacy and confidentiality of genetic information: what rules for the new science?," *Annu Rev Genomics Hum Genet* 2, 401-33.

Bjorkman, B. and Hansson, S.O. (2006), "Bodily rights and property rights," *J Med Ethics* 32:4, 209-14.

Cambon-Thomsen, A. (2003), "Assessing the impact of biobanks," *Nat Genet* 34:1, 25-6.

Cambon-Thomsen, A., Rial-Sebbag, E. et al. (2007), "Trends in ethical and legal frameworks for the use of human biobanks," *Eur Respir J* 30:2, 373-82.

Chadwick, R. and Berg, K. (2001), "Solidarity and equity: new ethical frameworks for genetic databases," *Nat Rev Genet* 2:4, 318-21.

Collins, F.S. and Green, E.D. et al. (2003), "A vision for the future of genomics research," *Nature* 422, 835-47.

Elger, B.S. and Caplan, A.L. (2006), "Consent and anonymization in research involving biobanks: differing terms and norms present serious barriers to an international framework," *EMBO Rep* 7:7, 661-66.

GIAN (2007), "Research project: 'Human Genetic Databases: Towards a Global Ethical Framework'" <http://www.ruig-gian.org/research/projects/project.php?ID=17>.

Gilbert, W. (1992), "A Vision of the Grail," in Kevles and Hood (ed.).

HapMap – The International HapMap Consortium (2005), "A haplotype map of the human genome," *Nature* 437, 1299-320.

Keller, E.F. (1995), *Refiguring Life: Metaphors of Twentieth-century Biology* (New York: Columbia University Press).

Kevles, D.J. and L.E. Hood (1992), *The Code of Codes: Scientific and Social Issues in the Human Genome Project* (Cambridge, Mass.: Harvard University Press).

Knoppers, B.M. (2005), "Of genomics and public health: Building public 'goods'?" *Cmaj* 173:10, 1185-6.

Knoppers, B.M. and Saginur, M. (2005), "The Babel of genetic data terminology," *Nat Biotechnology* 23:8, 925-7.

Kohane, I.S. and Mandl, K.D. et al. (2007), "Medicine. Reestablishing the researcher-patient compact," *Science* 316, 836-7.

Lin, Z. and Owen, A.B. et al. (2004), "Genetics. Genomic research and human subject privacy," *Science* 305, 183.

Lipworth, W. and Ankeny, R. et al. (2006), "Consent in crisis: the need to reconceptualize consent to tissue banking research," *Intern Med J* 36:2, 124-8.

Marshall, E. (2001), "Bermuda rules: community spirit, with teeth," *Science* 291, 1192.

Marshall, E. (2007), "Genetics. Sequencers of a famous genome confront privacy issues," *Science* 315, 1780.

Maschke, K.J. (2005), "Navigating an ethical patchwork—human gene banks," *Nat Biotechnology* 23:5, 539-45.

McFleming, J. (2007), "The governance of human genetic research databases in mental health research," *Int J Law Psychiatry* 30:3, 182-90.

McGuire, A.L. and Gibbs, R.A. (2006), "Meeting the growing demands of genetic research," *J Law Med Ethics* 34:4, 809-12.

Petsko, G.A. (2002), "No place like Ome," *Genome Biology* 3, 7 (comment 1010).

Sulston, J. and Ferry, G. (2002), *The Common Thread: A Story of Science, Politics, Ethics, and the Human Genome* (Washington, D.C.: Joseph Henry Press).

Tutton, R. and Kaye, J. et al. (2004), "Governing UK Biobank: the importance of ensuring public trust," *Trends Biotechnology* 22:6, 284-5.

PART I
RESEARCH BIOBANKS: CURRENT STATUS AND DEBATES

Chapter 2

Biobanks in the Literature

Bartha Maria Knoppers and Ma'n H. Abdul-Rahman[1]

1. Introduction

Population genomics uses basic data on genomic variation and on lifestyle behaviours and environmental factors to increase our understanding of common disease risk and human health. To carry out the needed studies of normal genomic variation across whole populations requires collecting biosamples and data on a longitudinal scale (Khoury 2004). Prior public consultation and engagement are a *sine qua non* for such activities. While the ethical frameworks and scientific norms by which existing or new database resources can be networked together are now being established (see especially the "Public Population Project in Genomics"), the literature is still replete with discussion of population biobanks.[2]

The literature on the ethical and regulatory aspects of human genetic databases reveals numerous outstanding issues; whereas once questions were raised about the very creation of biobanks, today three issues predominate—consent to the use of the samples, confidentiality, and potential commercialization—which are the subjects of the next three subparts of this chapter. The laws and policies governing biobanks have been examined elsewhere (Knoppers, Abdul-Rahman and Bédard 2007); the goal of this literature review is rather to discern whether these three areas of discussion in the literature reflect a gradual acceptance of the "legitimacy" of the approaches taken by the population biobanks or whether outstanding issues remain.

2. Consent

Consent of participants is a core issue since two interests are perceived as being at odds: 1) the protection of data, in light of the power of informatic technologies, and

1 The authors wish to thank Adam Spiro for his assistance and to acknowledge funding from Genome Quebec and Genome Canada and P3G (Public Population Project in Genomics).(www.p3g.org)

2 Population biobanks are "collections of biological materials having the following characteristics: i. the collection has a population basis; ii. it is established, or has been converted, to supply biological materials or data derived therefrom for multiple future research projects; iii. it contains biological materials and associate personal data, which may include or be linked to genealogical, medical and lifestyle data and which may be regularly updated; [and] iv. it receives and supplies materials in an organized manner" (Council of Europe 2006, Art. 17).

2) the importance of data-sharing and access to genetic resources by the scientific community (Uranga et al. 2005). Indeed, "informed consent is one part of honoring the contribution that the person is making to [the] advancement of knowledge" (Clayton 2005, 19). Informed consent to research includes information about the purpose, methods, risks and benefits of the study (Lipworth, Ankeny and Kerridge 2006; Knoppers 2005a), security and access policies, future uses and commercialization.

Population biobanks have unique features that are not fully addressed by the standard rules for disclosure to research subjects. For example, Article 22 of the Declaration of Helsinki requires that consent be specific to a clearly defined research project:

> In any research on human beings, each potential subject must be adequately informed of the aims, methods ... the anticipated benefits and potential risks of the study and the discomfort it may entail (World Medical Association 2000).

Yet in population biobanks, unlike clinical trials of drugs or devices, specific future research uses cannot be identified at the time of consent (Knoppers and Kent 2006; Gibbons et al. 2005). This central issue has been extensively debated in the literature, along with two associated issues, namely, the possible duty to recontact subjects about specific studies using their samples and related data, and the right of subjects to withdraw samples and data from the biobank.

2.1. Future uses and recontact

At the outset, it could be argued that the requirement of a specific consent is met even in these infrastructures. Their epidemiological objectives can be described in consent documents as can their longitudinal nature; the manner in which tissues will be conserved, the mechanisms for the security of data, and the ongoing governance structures for access and ethics monitoring can likewise be set forth when consent is obtained. In other words, specific consent is given for the creation of a public resource to be used for future research subject to these conditions. If a competent adult decides that these conditions and protections are sufficient, why would such a consent—broad as to future studies yet also specific as regards the biobank itself—not be valid? Two responses appear in the literature; the first is supportive of the validity of broad consent, while the second recommends "layered consent."

Some authors support the position taken by the national ethics committees of France and Belgium and by the HUGO Ethics Committee, endorsing the use of broad consent in this context (Gibbons et al. 2005). The authors maintain that under a broad consent regime, research participants may agree to future uses, including uses that are unforeseen at the time; the groups recognize, however, that such a broad consent has not been legally tested within Europe (Gibbons et al. 2005). The same is true in the United States where authors have approved the 1999 recommendations of the National Bioethics Advisory Commission that a valid consent for all the possible uses can be obtained at one time and further consent would not be required provided that certain restrictions were met (Trouet 2003, 417). Obviously, proper governance and ethics approval would have to be in place.

Others suggest "layering the consent." Under this approach, a participant could make choices, agreeing to allow a sample to be used in research on a specific disease or requesting recontact for consent to any other research (Trouet 2003, 411). In fact, open-ended consent may not be acceptable to participants who do not want their sample to be used for certain types of research (Lipworth, Ankenny and Kerridge 2006). Examples of potentially objectionable research include studies involving psychiatric diseases or sexual disorders or research of a commercial nature (Lipworth, Ankenny and Kerridge 2006; Trouet 2003).

It could be argued however, that contrary to a disease study, building a biobank for a large population study cannot be run with "individual" options. Each participant would have to be contacted again and again with additional information for each new research project. Moreover, samples shared with other researchers would have to be labeled as to the type of research or the choice of diseases allowed (Lipworth, Ankenny and Kerridge 2006; Trouet 2003). Both requirements would be extremely onerous and impractical in longitudinal studies of large populations. Indeed, in addition to being unable to get new consent from donors who have died, researchers would face difficulties in recontacting many living donors, such as those who have moved or who consider further contacts an invasion of their privacy or offensive if it reminds them of a previous illness (Lipworth, Ankeny and Kerridge 2006). Finally, the possibility of research collaborations would be greatly narrowed under such an approach.

In contrast, a one-time general consent "increases the scientific and social value of donated samples and lowers the costs of conducting research on them, eliminating the need to track the choices for each sample" (Lipworth, Ankeny and Kerridge 2006). One-time consent allows for data-sharing and prevents constant recontact for new consent. This approach is strengthened by recent studies examining different consent options, in which 75-95% of the respondents were keen to give one-time general consent, while at the same time relying on ethics committees to decide on the use of their samples (Wendler 2006). At a minimum however, population studies should inform the public on a regular basis of the nature of ongoing research using the biobank. Furthermore, population studies may wish to update data and so should ask at the time of recruitment for permission to recontact participants for this purpose. Such recontact would constitute a tacit renewal of consent and provides a renewed opportunity to withdraw.

2.2 Right to withdraw

The Nuremberg Code explicitly established the right to withdraw (U.S. G.P.O. 1949-1953, Art. 9), a liberty interest upheld in the Helsinki Declaration (World Medical Association 2000, Art. B-22). Interestingly, the literature on biobanks is not unanimous on the extent of this right in the context of population genetic studies or on when it may be exercised. Indeed, participants are often referred to as "donors," from which some commentators argue that as a "gift," samples and data cannot be returned or destroyed at the will of the donor, once the gift has been transferred to the recipient biobank (*Washington Univ. v. Catalona* 2007). Furthermore, the absence of the potential for physical harm to the participant—and considering that in contrast to clinical trials of new drugs or vaccine, studies of biological samples provided by biobanks are considered minimal risk—the "right" has less meaning in this context (Helgesson and Johnsson 2005). Yet, the majority of authors would uphold the right to withdraw arguing that the corollary of a broad consent would naturally entail the right to withdraw over time considering the trust inherent in such longitudinal studies (Helgesson and Johnsson 2005).

Irrespective, there is a remarkable absence of discussion on the practical aspects of the exercise of the right. Data in population studies is longitudinal, aggregated and largely epidemiological in nature. The number of samples needed for statistical significance is enormous and on the whole there are no individual, clinically validated results. To that end, theoretically "withdrawal" can take several forms: destruction of remaining samples and no further use of data; anonymization of the samples and data but allowing future use; and recall of any remaining samples including those that were transferred to outside researchers as well as destruction of all data not already incorporated into datasets. This multiplicity of options may ultimately prove to be impractical in this context due to the sheer size and often indefinite length of population genomic studies. At the level of whole populations, it may be wise simply to state that data and samples will no longer be used in the future unless already published or in aggregated datasets. Indeed, for all intents and purposes, it is the protection of the confidentiality of the samples that in fact serves to counterbalance a broad consent.

3. Confidentiality

While unanimous on the need to protect confidentiality and put in place security measures, no field is more rife with terminological confusion and philosophical differences than that of the confidentiality of data; and the mechanism chosen, such as coding or anonymization, obviously bears directly on whether it will be possible to communicate the results of any genetic analysis back to the people who supplied the biological samples.

3.1 Coding and anonymization

All population studies have in place procedures to ensure the confidentiality of data, balancing possible "information" risks of identification of the participant against the need to have data useful for quality research (Uranga et al. 2005). A related issue (although not under discussion here) is that of access by third party researchers to protected data and samples (Knoppers, Abdul-Rahman and Bédard 2007).

Coding is a well-known and increasingly sophisticated mechanism involving encryption, bar codes, simple, double or triple codes. What defines it is the ability to retrace the participant. In contrast, anonymization entails the irreversible breaking of any possible links back to the participant, though data available at the time of anonymization can accompany the sample. These two terms however have not been used or adopted in a clear and coherent fashion (Knoppers and Saginur 2005; see however ICH 2006). This is not without repercussions as the terms used determine not only possible future research uses by the study but by other researchers, that is, both the efficacy and utility of the study (Elger and Caplan 2006).

Coding permits both the constant downloading of data during the life of a participant (such as current health status, medical test results, etc.) and the return of results by researchers to a keyholder in a way that enriches the biobank for further research. Like in all information systems however, there can be no absolute certainty against potential breach of confidentiality.

Anonymization is ethically and legally expedient in the sense that no participant is traceable. Hence, a decade ago this was often the preferred route (Knoppers 2005b). Nevertheless, it greatly decreases the scientific and clinical utility of such initiatives as the data is static and not updated or dynamic. No rights can be violated because for all intents and purposes the person as an identifiable individual no longer exists. Certainly, no longitudinal study could function with anonymized data unless other highly informative databases provided demographic, administrative and other general, health data allowing perhaps for regional or sub-population profiling of some limited import. Where the current preference for coding may however turn into difficulty is as concerns the communication of results.

3.2 Communication of results

The return of results has long been limited to the notion of publication and more recently, for clinical trials, to international registries (for example, WHO 2007). Obviously, in clinical trials involving drugs or devices the return of results is the norm. The return of individual results in genetic research and hence in population genomic biobanks as well is, however, more problematic (Wilfond and Ravitsky 2006; Knoppers, Joly et al. 2006).

The literature on population biobanks is divided on this issue. The policy of "no-return" is supported by for example, the UK biobank (http://www.ukbiobank.ac.uk/) and CARTaGENE in Quebec (http://www.cartagene.qc.ca/). Certain authors argue that such initiatives, which build infrastructures that are largely epidemiological in nature, are not themselves involved in diagnostic tests of clinical validity and so have nothing to report (Knoppers, Abdul-Rahman and Bédard 2007; Knoppers, Joly et al. 2006). The complexity, costs, impracticality and implications of attempting to do so are self-evident and could compromise the security mechanisms put in place and so endanger confidentiality and even lead to discrimination. It could also be argued that it is misleading and undue inducement to hold out the promise of an eventual return of individual results.

Others argue that, respect for autonomy mandates return of results (Roberston 2003), even if they are only research results which by their very nature would be meaningless for the individual and unreliable (Buchanan, Califano, Kahn, et al. 2002). Moreover, even if reliable, "the consent form should state who will make the determination of reliability, according to what standards, and who will have the responsibility of informing the subject ... Other persons might find that always providing a subject that option is too costly to implement, but that recontacting with useful information should be pursued whenever feasible" (Roberston 2003, 305).

Nevertheless, considering the fact that some biobanks (e.g., Estonia) obtain ongoing access to the medical records of participants, the issue of individual feedback is real. If tests exist to validate research findings that may be clinically significant and treatment or prevention is available, procedures should be put in place for such exceptional circumstances (Johnston and Kaye 2004). In any event, the literature is unanimous that all population studies should have in place an ongoing public communication strategy not only while the resource is being built but also on the types of research protocols that use the data and samples (Robertson 2003). The presentation of the choice whether to receive information about individual genetic analysis or at a minimum, notification of the biobanks' policy in this regard, should be clear in the consent process. The same holds for eventual commercialization.

4. Commercialization

The debate on the characterization of genetic information—"property" or "person"— is classic (Chadwick 2001). It has also spilled over into the discussion of privacy which could be interpreted as a liberty interest or as a right of control (Rothstein 1997). Samples or data then are seen as either a proprietary right (Spinello 2004; Rule and Hunter 1999) or as an extension of personality rights (Le Bris and Knoppers 1997) (see 4.1). Neither seems to affect the issue of benefit-sharing following commercialization (see 4.2).

4.1 Ownership of samples

Twenty years ago, discussion in the literature largely stemmed from the infamous Moore case in California where unauthorized research uses of bodily materials led to a property claim by the donor (*Moore v. Regents of California* 1990; *cert denied* 1991). The economic inefficiency of such an approach to say nothing of the impact on altruistic donation of other elements of the body and organs was self-evident to the court. While Mr. Moore lost his property claim, the principle of obtaining an informed consent to research uses of samples and genetic information—and, in particular, the requirement that researchers reveal their proprietary or commercial interests—was established. Such a requirement exists irrespective of the "property" or "person" characterization and allows a research participant to exercise a right of control. The recognition of property rights (and so potentially patent rights) in the hands of the people who provide samples for research could also create obstacles for biomedical research. Indeed, the recognition of upstream property rights might not only impede downstream therapeutic applications but lead to multiple owners with ensuing licensing obligations and fees, etc. (Boyle 2003).

In short, a property approach is legally unworkable. Furthermore, while not denying the possibility of eventual intellectual property, some countries such as France have explicitly denied the possibility of "patrimonial" interests in the human body or its constituent parts. Whether the right of property or the right to personal autonomy is invoked, what is important then is the exercise of control. It should be noted however, that in the case of a "gift," ownership passes upon donation

(*Washington Univ. v. Catalona* 2007). If samples and data are provided altruistically for research and a test or a product is developed, is that the "benefit" that donors derive from the "gift" of their DNA?

4.2 Benefit-sharing

The concept of benefit-sharing dates back to the Rio *Convention on Biodiversity* of 1992. Having forsaken the common heritage of humanity approach for that of State Sovereignty over biological resources (plant and animal), balance was sought through the Bonn Guidelines that ensured the concept of benefit-sharing. In contrast, while upholding benefit-sharing, the Human Genome Project has largely adopted the common heritage approach. Thus, while undue inducement through compensation was forsworn, this did not mean that recognition and gratitude for the altruistic participation of citizens through benefit-sharing was not foreseen (HUGO 1996 and 2000).

The literature has largely seen this as a reasonable approach (Andrews 2005). Obviously, the notification of commercialization in the consent process warns participants that they will not individually share in any eventual profits and so "no ethical issues arise" (Roberston 2003). This may hold true for the large population genomic studies that create infrastructures for eventual research. Such projects are generally conducted for population health surveillance and research and only secondly, for profit-oriented research by industry (with the exception of deCode in Iceland) (Williams and Schroeder 2004). The "benefits" here are at the level of the population in terms of improving health care systems and providing a database for validation and replication to ensure better quality of science and health care generally.

In contrast, when companies collect samples and data for disease-gene hunting, "some patient or family groups may be unwilling to cooperate in setting up a biobank or research archive unless they have rights of access to final products, to licence patents, or even to a share in royalties" (Robertson 2003). Patent-sharing by families or groups may lead to lower costs for tests but could also in the long run create the same monopolies as industry would do in the case of exclusive licensing. To that end, a balance needs to be struck where the altruism of participants is "matched by a moral responsibility to use the resource, at least in part for the common good" (Williams and Schroeder 2004, 97). Benefit-sharing begins with the mechanism used to recognize, in an equitable way, the solidarity of citizens (Chadwick 2001). For companies, this can take the form first suggested by HUGO in 1996 and again in 2000, but for the population databases, the "benefit" would be that such primary data could be considered to be part of the public domain so as to enhance research opportunities (HUGO 2002).

5. Conclusion

Although it concentrates on the three main topics currently being debated—consent, confidentiality, and commercialization—this review of the literature on population biobanks reveals convergent as well as divergent positions.

There is convergence in the sense that the same issues are identified as require clarification. Thus, in the domain of consent, the question of the legitimacy of broad consent is now accompanied by the ancillary issues of the need to recontact for future uses and the exercise of the right to withdraw. The former is often subsumed in the broad consent question while the latter is upheld albeit without much specification as to the modalities by which it would be exercised. Likewise, discussion about the protection of confidentiality centres on traceability through coding or the lack thereof with anonymization. But the modalities of communication of results remain unresolved. Moreover, on this issue there is a certain confusion with the norms governing the return of results in clinical trials. At the outset there are no "individual" results in longitudinal population studies which by their very nature are not concerned with individuals and not organized to produce personal results. That being said, there is no doubt that over time such results may become possible.

The area where there is the most agreement is on the need to notify participants that commercialization may eventually arise as a result of researchers accessing a biobank for specific protocols. As such, donors of biological samples relinquish any "property" rights they may have in their material. Interestingly, the debate on gene patents has not entered the population biobank literature. This is probably due to the fact that such resources are not gene-hunting endeavours but epidemiological in nature. Nevertheless, in spite of their non-commercial nature, the concept of benefit-sharing is applied largely due to the tremendous public investment in such infrastructures. Thus researchers accessing these infrastructures are usually required to return their results to the biobank so as to not only enrich the data for use by others but also to ensure the quality of the data therein in order to speed potential use for population health.

The literature in these three areas is often ahead of the policies, the latter resulting from necessary compromises and consensus and are sometimes being drafted by individuals with little contact with the practical world of biobanks to say nothing of population genomics. In short, the need to move away from relying solely on individualistic ethics is evident in the literature on population studies which have communitarian goals (Knoppers and Chadwick 2005). The need to eschew anonymization but strengthen security and access mechanisms is self-evident given the longitudinal nature of such studies. The need to recognize and ensure access by researchers to these very public endeavours is crucial to their utility and success. Public investment, public participation and public trust demand no less.

Bibliography

Andrews, L.B. (2005), "Harnessing the benefits of biobanks," *Journal of Law, Medicine and Ethics* 33:1, 22-30.
Boyle, J. (2003), "Enclosing the Genome: What the Squabbles over Genetic Patents Could Teach Us," <https://www.law.duke.edu/boylesite/low/genome.pdf> under a Creative Commons License <http://www.creativecommons.org/licenses/by-sa/1.0>.

Buchanan, A., Califano, A., Kahn J. et al. (2002), "Pharmacogenetics: Ethics and Regulatory Issues in Research and Clinical Practice. Report of the Consortium on Pharmacogenetics, Findings and Recommendations," <http://www.utexas.edu/law/faculty/jrobertson/cv_jar1.pdf>, accessed 5 July 2007.

Chadwick, R. (2001), *Informed Consent and Genetic Research* (London: British Medical Journal Books).

Clayton, E.W. (2005), "Informed consent and biobanks," *Journal of Law, Medicine and Ethics* 33:1, 15-22.

Convention on Biological Diversity, signed 5 June 1992, U.N.T.S. No. 30619 (entered into force 29 December 1993).

Council of Europe (2006), *Recommendation Rec (2006)4 of the Committee of Ministers to Member States on Research on Biological Materials of Human Origin* (adopted 15 March 2006).

Elger, B.S. and Caplan, A.L. (2006), "Consent and anonymization in research involving biobanks," *EMBO Report* 7:7, 661-66.

Gibbons, S.M.C., Helgason, H.H., Kaye, J., Nomper, A., Wendel L. (2005), "Lessons from European population genetic databases: Comparing the law in Estonia, Iceland, Sweden and the United Kingdom," *European Journal of Health Law* 12:2, 103-134.

Helgesson, G. and Johnsson, L. (2005), "The right to withdraw consent to research on biobank samples," *Medicine, Health Care and Philosophy* 8:3, 315-321.

HUGO [Human Genome Organisation] Ethics Committee (2002), *Statement on Human Genomic Databases*, Human Genome Organisation, 1 December 2002.

HUGO [Human Genome Organisation] Ethics Committee (2000), *Statement on Benefit-Sharing*, Human Genome Organisation, 9 April 2000.

HUGO [Human Genome Organisation] Ethics Committee (1996), *Statement on the Principled Conduct of Genetics Research*, Human Genome Organisation, 31 March 1996.

International Conference on Harmonization of Technical Requirements for Registration of Pharmaceuticals for Human Use [ICH] (2006), "Final Concept Paper E15: Terminology in Pharmacogenomics," endorsed by the ICH Steering Committee 19 April 2006, online <http://www.ich.org/LOB/media/MEDIA3099.pdf>.

Johnston, C. and Kaye, J. (2004), "Does the UK Biobank have a legal obligation to feedback individual findings to participants?," *Medical Law Review* 12:3, 242-243.

Khoury, M.J. (2004), "The case for a global human genome epidemiology initiative," *Nature Genetics* 36:10, 1027-28.

Knoppers, B.M., Abdul-Rahman, Ma'n H. and Bédard, K. (2007), "Genomic databases and international collaboration," *KLJ* 18:2, 291-311.

Knoppers, B.M., Joly, Y., Simard, J., Durocher, F. (2006), "The emergence of an ethical duty to disclose genetic research results: international perspectives," *European Journal of Human Genetics* 14, 1170-78.

Knoppers, B.M. and Chadwick, R. (1994), "Human Genome Project: Under an International Legal and Ethical Microscope," 265 *Science*, 2035-36.

Knoppers, B.M. and Kent, A. (2006), "Ethics watch: policy barriers in coherent population-based research," *Nature Reviews Genetics* 7:1, 8.

Knoppers, B.M. (2005a), "Consent Revisited: Points to Consider," *Health Law Review* 13:2-3, 33-8.

Knoppers, B.M. (2005), "Biobanking: international norms," *The Journal of Law, Medicine and Ethics* 33:1, 7-14.

Knoppers, B.M. and Chadwick, R. (2005), "Human genetic research: emerging trends in ethics," *Nature Reviews: Genetics* 6:1, 75-9.

Knoppers, B.M. and Saginur, M. (2005), "The Babel of Genetic Data Terminology," *Nature Biotechnology* 23, 925-27.

Le Bris, S. and Knoppers, B.M. (1997), "International and Comparative Concepts of Privacy," in M. Rothstein (ed.), *Genetic Secrets* (New Haven: Yale University Press), 418-48.

Lipworth, W., Ankeny, R. and Kerridge, I. (2006), "Consent in crisis: the need to reconceptualize consent to tissue banking research," *Internal Medicine Journal* 36:2, 124-28.

Moore v. *Regents of California* (1990) 793 F 2d 479 [Cal.]; *cert denied* (1991) 111 S. Ct. 1388.

Nuremberg Code, from Trials of War Criminals before the Nuremberg Military Tribunals under Control Council Law No. 10. Nuremberg, October 1946-April 1949. Washington, D.C.: U.S. G.P.O, 1949-53.

Public Project in Population Genomics (P3G) n.d., <http://www.p3gconsortium.org>.

Robertson, J.A. (2003), "Ethical and Legal Issues in Genetic Biobanking," in B.M. Knoppers (ed.), *Populations and Genetics: Legal and Socio-Ethical Perspectives* (Leiden/Boston: Martinus Nijhoff Publishers, Netherlands), 297-309.

Rothstein, M. (1997), *Genetic Secret* (New Haven: Yale University Press).

Rule, J. and Hunter, L. (1999), "Towards Property Rights in Personal Data," in C. Bennet and R. Grant (eds), *Visions of Privacy* (Toronto: University of Toronto Press), 168.

Spinello, R.A. (2004), "Property rights in genetic information," *Ethics and Information Technology* 6:1, 29-42.

Trouet, C. (2003), "Informed consent for the research use of human biological materials," *Medicine and Law* 22:3, 411-19.

Uranga, A.M. et al. (2005), *Outstanding Legal and Ethical Issues on Biobanks: An Overview on the Regulations of Member States of the EuroBioBank Project* (Madrid: Instituto de Salud Carlos III).

Washington Univ. v. Catalona, United States Court of Appeals, June 2007 (No. 06-2286).

Wendler, D. (2006), "One-time general consent for research on biological samples," *BMJ* 332, 544-47.

WHO International Clinical Trials Registry Platform (2007), World Health Organization, http://www.who.int/ictrp/en/.

Wilfond, B.S. and Ravitsky, V. (2006), "Disclosing individual genetic results to research participants," *The American Journal of Bioethics* 6:6, 8-17.

Williams, G. and Schroeder, D. (2004), "Human genetic banking: altruism, benefit and consent," *New Genetics and Society* 23:1, 89-103.

World Medical Association (2000), *Declaration of Helsinki*, adopted by the 52nd World Medical Association General Assembly, Edinburgh, Scotland.

Chapter 3

Guidelines on Biobanks: Emerging Consensus and Unresolved Controversies

Effy Vayena, Agomoni Ganguli-Mitra and Nikola Biller-Andorno

1. Introduction

The ever increasing importance of human biological samples in biomedical research has made biobanks[1] a central and inextricable element of biomedical progress. The fact that biobanks promise research that will bring science and medicine closer to a deeper understanding of health and disease has led to a rapid growth of public, commercial, international, and national biobanks with millions of samples already stored (Kaiser 2002).

A parallel growth, although not proportional, has been seen in development of international, national and professional guidelines, recommendations and even legislation attempting to address the many ethico-legal issues that arise in biobanking. While many of these guidelines use similar approaches, stemming from widely accepted principles, they still use different ethical frameworks and therefore present significant differences, controversial recommendations and leave a number of issues unresolved (Knoppers 2005). The focus of this chapter is on international guidelines and particularly on how international recommendations address the key issues of consent, anonymization, access, ownership and benefit-sharing.

This review includes the following international guidelines that address issues related to biobanks: the Human Genome Organisation Ethics Committee's *Statement on Human Genomic Databases* of 2002 (hereafter HUGO), the United Nations Educational, Scientific and Cultural Organization's *International Declaration on Human Genetic Data* of 2003 (hereafter UNESCO), the World Health Organization's report entitled *Genetic Databases - Assessing the Benefits and the Impact on Human and Patients Rights* of 2003 (thereafter WHO), the World Medical Association's *Declaration on Ethical Considerations Regarding Health Databases* in 2002 (hereafter WMA). In addition we consider the recent Council of Europe's *Recommendation RecRE(2006)4 of the Committee of Ministers to Member States on Research on Biological Materials of Human Origin* (hereafter COE) as it explicitly refers to biobanks and is intended for use in many countries. We have also included the US National Bioethics Advisory Committee's recommendations (hereafter NBAC) as they have been followed by the OHRP and as such are relevant

---

1 For the definition and use of the terms "biobank" and "human genetic databases" see chapter 1.

for the large number of biobanks that are in operation in the United States as well as for international projects with US participation. They also form the normative framework for one of the samples of our study.[2] We have not included any other national legislation or professional society guidelines, although we do use some examples from both types of regulations when they highlight extreme differences in position about the key issues.

This chapter shows that international instruments still lack harmonization while they fail to address several critical issues. As a result, national policies are developed with variations that can be problematic for international research since what is acceptable and legal in one country is not necessarily so in another (Maschke and Murray 2004, Pearson 2004). With international research becoming increasingly popular and important, an international ethical framework, addressing all the topics that are relevant for biobanking in the international context has become a matter of urgency (Elger and Caplan 2006; Kaye 2006).

2. Informed Consent

Respect for autonomy is one of the basic ethical principles which dictates the requirement for informed consent for any potential participant. The right to freely and voluntarily decide whether one wants to participate in research after having been given all relevant information is widely recognized by international guidelines and national laws regulating research on human subjects. Precise language indicating the free, informed and voluntary participation is almost standard in consent form documents and for the majority of research projects such a document is signed by the individual participant, thumb-printed by an illiterate participant, or signed by a proxy or legal guardian in case the participant is a minor or incapable of giving valid consent. Providing biological samples for research purposes is also considered participation in research, therefore the consent requirement also applies to obtaining human samples for biobanking. However, the long-term storage of biological samples and data raises issues that are not addressed by one-time consent procedures as used in traditional research settings.

2.1 Types of consent for biobanks

According to the Council for International Organizations of Medical Sciences (CIOMS)'s *International Ethical Guidelines for Biomedical Research involving Human Subjects* (the international standard for biomedical research) and many other guidelines, before an individual is asked to become a research participant, they should be given essential information on which to base the decision. In Guideline 5, CIOMS lists twenty-six essential items that should be included in the consent document, all of them related to the specific research protocol. In the case of biobanks, samples may be collected for use in a particular research project, various closely related projects, or future unforeseen projects. Depending on what they are collected for

---

2 Cf. chapter 4.

and how they are going to be used, the detail of available information to the sample donor may be limited. Does limited information about research details invalidate consent as one would argue it would do in the case of other biomedical research projects? If the sample were to be used for a project different to the one described in the consent form would the individual have to re-consent? What if the individual cannot be reached? International guidelines do not have a unanimous answer to these questions and consequently a variety of recommendations as regards valid consent has emerged.

The HUGO recommendation for example includes several options: "consent may include notification of uses (actual or future), or opting out, or in some cases, blanket consent." This leaves the consent procedure adaptable with regard to different types of research projects without, interestingly enough, even suggesting that researchers should, where possible, comply with more restrictive procedure. UNESCO is slightly stricter in suggesting that consent should be obtained for "collection ... subsequent processing, use and storage" of samples (UNESCO, Art. 8) without exactly specifying the degree of detail that should be included. In later articles it suggests that samples should not be used for "a different purpose that is incompatible with the original consent" unless an important public health issue is at stake (UNESCO, Art. 16). However, this recommendation leaves the definition of "incompatibility" open to interpretation (Knoppers 2005).

WHO's recommendation, although more specific than the HUGO recommendation in calling for the incorporation of information on future use of data, allows for blanket consent for future research provided that anonymity can be guaranteed. It is perhaps worth noting that the WHO's definition of "anonymity" is different from the term "anonymization" as used by HUGO (see Section 3.2 on terminology). However, WHO's report (WHO, Rec. 14) even considers it permissible to "depart from the practice of requiring active informed consent prior to participation in the creation of a genetic database" if all of the following criteria are met: a) a significant public health issue is identified, b) educational programs are instituted and public debate is possible, c) privacy is protected, d) anyone can refuse to participate, e) the process is ethically scrutinized. This position substantially departs from the traditional notion of informed consent. It implies a shift from the individual responsibility for a fully informed decision to a collective (researchers, community, ethics review committee etc.) responsibility for such a decision but with mechanisms in place that should serve the individual's best interest.

The recent document by the Council of Europe calls for consent "as specific as possible with regard to any foreseen research uses" (COE 2006). It also specifies that if a research project is beyond the scope of the original consent then re-consent is necessary. If it is not possible to locate the individual for re-consent it still allows the use of the biological materials if certain conditions (important scientific interest, no other way of obtaining the information and no evidence that the individual would have been opposed to such use) are met. WHO also recommends the use of archival material after anonymization (WHO, Rec. 10). In the case of WMA, as participants are not only considered as such, but also as patients and databases as containing personal health information, all matters regarding consent, control and access to data are more strictly focused on participant (patient) rights. In general, participants must be notified regarding all use of their data, unless there are compelling reasons to act otherwise (WMA, Principles 11, 16-19). Interestingly, differing from other guidelines, in the case of research that had not been "envisaged at the time the data were collected," it recommends the decision of whether patients need to be contacted for re-consent be left to an ethical review committee (WMA 2002, Principle 21).

Finally, the NBAC proposes the so-called multi-layered consent where the individual has the option to make a number of choices such as permitting only one type of research, all types of research, or agreeing to be called for re-consent etc. (NBAC 1999, Elger and Caplan 2006). The multi-layered consent essentially makes the individual the ultimate decision-maker. However, the document, similarly to others, includes provisions for proceeding without consent if certain criteria are met.

2.2 Group/collective consent

International guidelines for research ethics consider informed consent as individual. In the case of traditional research settings, the before mentioned CIOMS guidelines and the Nuffield Council's report on research explicitly state that a group can authorize a research project by permitting the research to be implemented but consent can only be given by the individual. (CIOMS 2002, Nuffield 2002 and 2005) When research results may carry risks for an entire group or community, for example the risk of stigmatization, it is justified and even advisable to ask the community for permission, but individual consent is needed for participation. In some types of research, in particular genetic research, studying samples obtained from consenting individuals may reveal private and confidential information for their families or extended relatives. Furthermore, research with consenting individuals from a specific community may reveal information for the entire community. The question that arises in such cases is whether individual consent is adequate and what measures should be taken by the investigators to protect not only the individual who provides the samples but also those who might be affected by his or her participation. Challenging the traditional notion of individual consent one professional society, the European Society for Human Genetics, has suggested that in population studies additional group consent may be required (ESHG 2003). The guidelines do not provide further explanations as regards the situations in which group consent may or may not be required, and it is not clear whether in some cases group consent might outweigh individual consent or vice versa. It is however difficult to envision how research ethics could move from the strict notion of individual and specific consent (a notion deeply rooted in the principle of autonomy and self-determination) for the purposes of biobank research while maintaining this standard for all other research.

Given that genetic data is by nature shared and that a large number of biobanks store genetic data, it is noticeable that international guidelines hardly address the issue of collective or group involvement. The HUGO guidelines mention the role of communities, stating that "the choice and privacy of individuals, families and communities should be respected" (HUGO, Rec. 4). However, this leaves open and undecided in what way or to what extent families and communities should be included in consent procedures and other decision-making. The WHO guidelines raise the issue of group consent (in the context of population studies) in the preamble to recommendation 14 and consider the alternative of assuming that the population in question consents unless individuals indicate otherwise, but it recognizes the controversial nature of such an approach.

2.3 Consent withdrawal

Most international guidelines agree on the principle that an individual maintains the right to withdraw consent at any point during a research project and that such action should not have any negative consequences for the person's medical care. However, the question here is what happens to the biological materials that are already stored in a biobank after consent is withdrawn and what happens to data that may already have been derived by the time the consent is withdrawn? UNESCO states that once a person withdraws consent "the person's genetic data, proteomic data and biological samples should no longer be used unless they are irretrievably unlinked to the person concerned" (UNESCO, Art. 9). It further suggests that if the data and samples are not irretrievably unlinked they should be destroyed. Similarly, COE (COE, Art. 15) and WHO (WHO, Rec. 18) allow for withdrawal of consent or alteration of the scope of consent and in such case the implication for the samples and data is that they are either to be rendered unlinked, anonymized (or absolutely anonymized in the case of WHO) or destroyed. While most guidelines seem to agree on this issue, they are silent on the practical aspects of implementing the withdrawal of consent such as in cases where results have already been obtained and incorporated into other databases, when samples are rare and the continuation of important projects might be jeopardized etc. As a result a number of national guidelines have emerged providing practical alternatives with variation. For example in Sweden the law states that "if the withdrawal of consent refers to all use" the samples must be destroyed or de-personalized (Sweden Chapter 3, Section 6) without specifying acceptable scenarios if withdrawal is not for "all use." In Iceland the law states that although samples should be destroyed if consent is withdrawn, research results obtained from samples for which consent has been withdrawn should not be destroyed (Iceland, Art. 7).

The 2002 HUGO guidelines are silent on the issue of withdrawal. All matters related to consent and use of samples are grouped and addressed under the obligation to respect "the choices and privacy of individuals, families and communities" (HUGO, Art. 4). However, it is worth noting that in the 2002 document HUGO specifically recalls in its preamble its 1998 statement entitled, *DNA Sampling: Control and Access*. In these earlier guidelines, HUGO stresses the shared nature of genetic data and recommends that at the request of a participant, stored samples may be destroyed unless there is need for access by other relatives and unless the data have already been "provided to other researchers or if already entered into a research protocol or used for diagnostic purposes" (HUGO 1998).

3. Protecting confidentiality

Biological samples contain potentially sensitive information about their participant. An important issue in the preparation of guidance for biobanks has been how to best protect such confidential information. The issue of protecting confidentiality can be framed within the principle of autonomy (recognizing participants' right to decide which information should be revealed and their expectation that this information will not be divulged to third parties in any way that is not consistent with what has been consented to) but also within the principle of beneficence/non-maleficence by recognizing the importance of avoiding harm, as sensitive information that is divulged to others may be used to the disadvantage of the individual.

3.1 Coding and anonymization

The most commonly suggested way to protect confidentiality in biobanks is by ensuring that the donor is not identifiable. The WHO guidelines describe two forms of anonymity: "*Absolute anonymity* ... when no means are available to link data to an identifiable individual" and "*proportional* or *reasonable anonymity* ... when no reasonable means of identification of specific individuals is available" (WHO preamble Rec. 7). In practice the latter is done by double coding, whereby numbers are assigned to participants and their samples but the code linking the number to personally identifying information is only accessible to a few authorized persons. Absolute anonymity is obtained by having identifiers linked to samples removed later on in the project; or for example by removing all identifying links already at the collection of sample (Swede et al. 2007). Current biobanks use some form of protection, ranging from minimal to complete anonymity, although the latter has been challenged as theoretically unachievable (Lin et al. 2004). However, there is certainly no uniformity in the application of various means of protection across existing biobanks.

All guidelines reviewed here address the importance of protection. However, specific information about the desirable degree of identifiability and how to achieve it is missing. UNESCO recommends that "human genetic data, human proteomic data and biological samples collected for the purposes of scientific research should not normally be linked to an identifiable person" but further on states that data and samples "collected for medical and scientific research purposes can remain linked to an identifiable person, only if necessary to carry out the research and provided that the privacy of the individual and the confidentiality of the data or biological samples concerned are protected in accordance with domestic law" (UNESCO, Art. 14). Similarly, the WMA guidelines propose that if de-identifying the data is not possible then the identity should be protected by an "alias or code" (WMA, Principle 24). The COE takes an even more flexible stand by recommending that "data should be anonymized as far as appropriate to the research activities concerned" but that the research would need to make appropriate justifications for the uses of data "in an identified, coded or linked anonymized form" (COE, Art. 8). The interesting proposition by COE is that anonymization should be overseen by a review process. Similarly, WHO suggests that "proportional" anonymity is acceptable but what really matters is how this is done, who has access to the data and what is included in the original consent. It concludes that anonymity alone without proper review might still be inadequate and recommends multiple tasks for the ethics review committee including: "a) to scrutinize and ensure the legitimacy of requests to the database; b) to act, where possible, as an intermediary between the creators and the users of the database, in respect of decoding apparatus used to anonymize and/or link data held on the database; c) to maintain standards and keep anonymization processes under review" (WHO, Para. 4.2, preamble to Rec. 7 and Rec. 7). The suggestion for a review process implies that the decision about anonymity would be made on a case by case basis. This approach would actually allow a better assessment of the risk-benefit-ratio for individual projects, but would also presuppose that ethics committees have sufficient competency in statistics and general data protection issues. HUGO, finally, recommends informing participants of the degree of identifiability of their data (HUGO, Rec. 4(c)).

In general while guidelines favor anonymization the caveats that are embedded in most of them reveal the understanding that complete anonymity could actually limit the research potential of biobanks. In addition complete anonymity would make recontacting the participant impossible and therefore it would exclude re-consent as well as informing the participant about important research results that might be of clinical value.

Nevertheless, it is unclear whether determining the procedural aspects on a case by case basis or leaving it up to national laws, in fact, facilitates or impedes the international movement of samples and data.

3.2 A remark on terminology

In discussing the various means of protecting confidentiality, guidelines use very different terms to describe the different degrees of anonymization of samples and data.

In the guidelines reviewed here the terms included are: "anonymized," "aggregate," "confidential," "coded," "completely anonymized," "de-identified," "irretrievably unlinked to identifiable person," "linked," "linked to identifiable person," "linked anonymized," "linkable coded," "unlinked to identifiable person," "personally identifiable", "proportional anonymity," "reasonable anonymity," "unlinked anonymized," "unlinked." Several studies that focused on the issue of terminology and reviewed national guidelines and legislation identified a plethora of terms. (Knoppers and Saginur 2005; Elger and Caplan 2006). Moreover, in many cases, the same terms are used to designate different procedures and levels of protection. To take a few examples: HUGO states that samples can be "coded, anonymized, aggregate etc." (HUGO, Rec. 4(c)). Here, "coded" can be interpreted as meaning that samples have been assigned a code, and the key to the code is kept safe. "Anonymized" is used to imply the destruction of such a code, breaking all (obvious) links between the participant and their sample. COE (COE, Art. 8), on the other hand, uses the term "coded" for coded samples where the researchers have access to the code, and "linked anonymized" in cases where the researcher does not have access to the code. Finally, WHO (WHO, Para. 4.2, preamble to Rec. 7), uses the terms "absolute anonymity" for cases where the participant-sample link has been destroyed, and "proportional or reasonable anonymity" where such a link still exists, albeit in a coded form. Although there are some trends in the terminology there is certainly no uniform language, which is confusing and can lead to miscommunication, especially in the context of international projects. Identifiability and its degrees have implications on privacy and confidentiality and consequently on determining the risk-benefit-ratio of a research project. Therefore clarity and precision in terminology is critically important particularly if a global framework is to be developed. Moreover, it has been suggested in recent literature that genetic data cannot be absolutely anonymous due to the unique nature of individuals' DNA sequences. It is therefore worth revising the terminology and protective mechanisms recommended in guidelines, particularly in the case of genetic databases (Lin et al. 2004).

3.3 Third party access to biobank data

Even if protection of data confidentiality is achieved, the issue persists of who should have access to the pool of samples and related information stored in biobanks and to what extent. Although many guidelines reiterate the right of participants to access their samples and data as well as the general prohibition to transfer information to third parties such as employers and insurance companies, the issue of access for research purposes remains debated. The HUGO guidelines, adhering to the organization's general approach to genetic databases as global public goods, encourage the "free flow of data" (HUGO, Rec. 3) as well as access to the benefits arising from this free sharing. Moreover, they indirectly address the possibility of territorial limitations (i.e. keeping genetic material within national borders), by specifying that "insofar as it benefits humanity, the free flow, access and exchange of data are essential. Cooperation and coordination between industrialized and developing countries should be facilitated" (HUGO, Rec. 3(a)). Similarly, UNESCO advises researchers to "establish cooperative relationships," and to "make every effort ... to continue fostering the international dissemination of scientific knowledge concerning human genetic data ...." However, UNESCO also recognizes that the cross-border flow of human genetic data should be in accordance with domestic laws (UNESCO, Art 18). Unfortunately, in the context of international research, the cross-border flow of genetic data may be distributed unequally as some countries have more restrictive laws regarding territorial limitations than others (Dickenson 2004, 117). The COE guidelines, while allowing access to biological materials by researchers (COE, Art. 20), also specify that in the case of cross-border flow of materials and associated personal data should "only be transferred to another state if that state ensures an adequate level of protection" (COE, Art. 16). Finally, of particular interest are also the WHO recommendations which state that "it should be the role of an independent body to oversee and regulate access to genetic databases" (WHO, Rec. 17). It further recommends that the same body should review the anonymization methods and the application of anyone who requests access, implying that access is a matter that needs to be regulated under close supervision and on a case by case basis.

4. Ownership of biological samples

Ownership of biological samples is a contested issue (Winickoff and Winickoff 2003). A variety of contradicting or vague recommendations and legislative approaches reflect the difficulty in agreeing on the ownership of biological samples once they have been removed from the human body. At the international level too, guidelines differ substantially.

HUGO's approach is that databases are public goods belonging to all and allowing all to benefit from the knowledge that will be derived; knowledge that eventually belongs to humanity. The HUGO statement, along with a similar statement by UNESCO, emphasizes the notion of common heritage (Knoppers 2005). Without directly contradicting this position, other international guidelines either attribute ownership to the individual who provides the samples or bestow some type of ownership to the biobank where the samples have been stored and processed. WHO for example suggests in recommendation 19 that "serious consideration should be given to recognizing property rights for individuals in their own body samples and genetic information derived from those samples" (WHO 2003). However, an earlier guideline by the Nuffield Council proposed that the user of the sample (the biobank) acquire possessory rights and possibly a right of ownership over the tissue once removed (Nuffield 1999). At the national level various versions of the above recommendations have been adopted, for example, giving the biobank "custodianship" in the UK or by recognizing a status of corporate shareholder for the institution and researcher involved in Brazil (Boggio et al. 2005).

The issue of ownership becomes even more complex in specific biobanks such as the umbilical cord blood biobanks. What is interesting in the case of cord biobanks is that ownership is not even clear before the cord is donated (Annas 1999). While the mother is assumed to be the one to consent, the blood is assumed to be owned by the baby. Who could decide about future use, who has the right to withdraw the sample, at which point in time is that permissible, etc.? All these questions are answered on an ad hoc basis. While there is an increasing number of such biobanks, there is barely any guidance at the international level.

4.1 Commercialization

Ownership of samples is an important question with serious implications on commercialization. The information derived by samples and its potential for profit-making applications raises the question of where such profits should go: to the sample donor, the biobank or another entity? The probability of high financial gains from potential applications makes the issue of ownership even more difficult to resolve. Another aspect of the same issue is whether samples can actually be sold once they have been obtained by a biobank. Should a biobank be allowed to sell samples to a third party and should in this case the individual donor benefit? How would remuneration be regulated? What would the sense of withdrawing consent mean if the biobank had the authority to sell?

Commercialization is a topic that international guidelines are reluctant to address directly. There are some general statements suggesting that human tissue and cells should not be commodities that generate financial gain. The *Convention on Human Rights and Biomedicine* of the Council of Europe (1997) included that "The human body and its parts shall not, as such, give rise to financial gain" but this was more intended to prohibit commerce and trafficking of human organs and body parts. The same position has been taken in the 2006 recommendations by the COE that are specifically addressing biobanking. However, these statements leave enough room for interpretation. First and foremost they do not distinguish between samples and data derived from sample analyses. Data may be of higher commercial interest than samples themselves. Related to commercialization is the issue of patenting, which few of the international guidelines address. Of interest are the WHO guidelines, which mention, not in a recommendation but in Paragraph 7.2, that intellectual property rights may be granted to researchers or those who create and manage genetic databases: "patent rights are available for inventions using human material, and both copyright and database rights can be claimed in respect of the structure, content, selection and arrangement of a database." However, the same paragraph also specifies that "such rights do not accord an unfettered reign to the rights holders to do what they wish with their property." WHO also does not cross out the commercial aspect of a biobank but specifies that participants should be notified of a biobank's "commercial potential" (WHO, Rec. 6(1)). On a similar note, the HUGO guidelines allow for researchers, institutions and commercial entities to have a "right to fair returns" (HUGO, Rec. 6), but also add that any fees should not restrict the flow of information or access.

5. Benefit-Sharing

As international biomedical research and scientific cooperation between nations with different levels of economic development increases, there has been a heightened call for sharing the benefits of research with participating individuals and populations, both as a matter of reciprocity and with regard to global justice. Defining the type and nature of benefit (direct, indirect, collective, individual, financial or other) as well as the exact recipient (group, community, population) has been the subject of much discussion and controversy (Schulz-Baldes et al. 2007).

Some biobank guidelines lay out recommendations for benefit-sharing, although this is certainly not widespread among all international instruments. Among those which address benefit-sharing as an independent topic, UNESCO's Article 19 recommends benefits to be shared with "society as a whole and the international community" and lists several ways to do so: (i) special assistance to the persons, and groups that have taken part in the research; (ii) access to medical care; (iii) provision of new diagnostics, facilities for new treatments or drugs stemming from the research; (iv) support for health services; (v) capacity building facilities for research purposes; (vi) development and strengthening of the capacity of developing countries to collect and process human genetic data, taking into consideration their specific problems; (vii) any other form consistent with the principles set out in this [the UNESCO] Declaration. WHO takes the stand that research should eventually benefit either the individual or the group to which the individual belongs without giving explicit examples.

While the HUGO guidelines are vague on the nature and extent of benefits that should be shared from research using genomic databases, it is important to note that the preamble to the recommendations mentions the 2000 HUGO Ethics Committee *Statement on Benefit Sharing*, which discusses this topic in detail, as well as how it should be regulated. While these recommendations address some of the most fundamental topics in benefit-sharing, such as the ethical principles behind benefit-sharing as well as practical questions, such as how to define a community, they also reflect the procedural difficulties in applying general guidelines to various types of research and in deciding how and to whom to attribute particular rights and obligations in specific research settings (HUGO 2000).

In general, even where they address benefit-sharing, international guidelines on biobanking and research involving human biological samples do not discuss the type and nature of benefits. They lack explicit recommendations on the many procedural issues relating to benefit-sharing such as defining the "community" or "group" that should benefit or a mechanism for benefit distribution. This is of particular interest as in many population studies and specifically population genetic studies the benefits are more likely to be collective than individual and it would therefore be useful to have more clarity on what such benefits should be. Furthermore the guidelines are completely silent on gender issues and the role that gender inequities can play in achieving fair distribution of benefits (Alvarez-Castillo and Feinholz 2006).

6. Conclusions

While the described set of international guidelines is of prime importance for biobanking they have been for the most part more general than explicit. Moreover, they are not necessarily consistent in their recommendations and when guidance is necessary they often delegate the decisional authority to national laws or to reviews to be carried out by research ethics committees. In some instances a case by case decision might be the best option. However the lack of a uniform framework within which a decision is made can easily result, as it already has, in various approaches and contradicting decisions about the same aspects of a research project. Furthermore, international guidelines do not adequately address important global issues, such as territorial limitations or collective consent. Some guidelines are notably distant from the reality of contexts they are supposed to regulate, and pay little attention to the particular needs of international research projects with biobank data.

Nevertheless, the existing documents provide a suitable starting point from which a well-composed global framework could be built. Such a framework would take up missing points, particularly those that spring from international collaborations, it would harmonize existing variation while leaving adequate room for differing national solutions, and it would be sufficiently concrete to be meaningfully applied to the medical, laboratory and business environments that shape biobanking.

Bibliography

Alvarez-Castillo, F. and Feinholz, D. (2006), "Women in developing countries and benefit sharing," *Developing World Bioethics* 6, 113-21.

Anderlink, M.R. and Rothstein M.A. (2001), "Privacy and confidentiality of genetic information: what rules for the new science?," *Ann Rev Gen and Hum Gen* 2, 401-33.

Annas G.J. (1999), "Waste and longing—the legal status of placental-blood banking," *N Eng. J Med* 340: 19, 1521-4.

Boggio, A., Biller-Andorno, N., Elger, B., Mauron, A. and Capron, A.M. (2005), "Comparing guidelines on biobanks: emerging consensus and unresolved controversies" <http://www.ruig-gian.org/research/outputs/output.php?ID=254>, accessed 17 September 2007.

Council of Europe (1997), Convention for the Protection of Human Rights and Dignity of the Human Being with Regard to the Application of Biology and Medicine: Convention on Human Rights and Biomedicine. CETS No. 164.

Council of Europe (2006), *Recommendation Rec(2006)4 of the Committee of Ministers to Member States on Research on Biological Materials of Human Origin* (Strasbourg, France: Council of Europe).

Council for International Organizations of Medical Sciences (2002) *International Ethical Guidelines for Biomedical Research Involving Human Subjects* (Geneva, Switzerland: CIOMS)

Dickenson, D. (2004), "Consent, commodification and benefit-sharing in genetic research," *Developing World Bioethics* 4: 2, 110-24.

Elger, B. and Caplan, A. (2006), "Consent and anonymization in research involving biobanks," *EMBO Reports* 7, 661-6.

European Society for Human Genetics (2003), "Data storage and DNA banking for biomedical research: technical, social and ethical issues. Recommendations of the European Society of Human Genetics," *European Journal of Human Genetics* 11, Suppl. 2, S8-10.

HUGO [Human Genome Organisation] Ethics Committee (1998), *Statement on DNA Sampling: Control and Access* <http://www.hugo-international.org/Statement_on_DNA_Sampling.htm>, accessed August 1 2007.

HUGO [Human Genome Organisation] Ethics Committee (2000) *Statement on Benefit-Sharing* <http://www.hugo-international.org/PDFs/benefit.html>, accessed September 18, 2007.

HUGO [Human Genome Organisation] Ethics Committee (2003) *Statement on Human Genomic Databases* <http://www.hugo-international.org/Statement_on_Human_Genomic_Databases.htm>, accessed 17 September 2007

Iceland (2000), Act on Biobanks no 110/2000 <http://www.personuvernd.is/tolvunefnd.nsf/pages/95EAE39BAC9DFA25002569050057034C>, accessed August 1, 2007.

Kaiser, J. (2002), "Population databases boom, from Iceland to the U.S.," *Science* 298, 1158-61.

Kaye, J. (2006), "Do we need a uniform regulatory system for biobanks across Europe?," *European Journal of Human Genetics* 14, 245-48.

Kegley, J.A. (2004), "Challenges to informed consent," *EMBO Reports* 5, 832-6.

Knoppers, B.M. (2005), "Biobanking: international norms," *J Law Med Ethics* 33, 7-14.

Knoppers B.M. and Saginur, M. (2005), "The Babel of genetic data terminology," *Nat Biotechnology* 23, 925-7.

Lin, Z., Owen, A.B. and Altman, R.B. (2004), "Genetics: Genomic research and human subject privacy," *Science* 305: 5681, 183.

Maschke, K.J. and Murray, T.H., (2004), "Ethical issues in tissue banking for research: the prospects and pitfalls of setting international standards," *Theor Med Bioeth.* 25: 2, 143-55.

National Bioethics Advisory Committee (1999) "Research Involving Human Biological Materials: Ethical Issues and Policy Guidance" <http://bioethics. georgetown.edu/nbac/hbm.pdf> accessed 19 September 2007.

Nuffield Council (1999), *Human Tissue: Ethical and Legal Issues* (London: Nuffield Council on Bioethics).

Nuffield Council (2002), *The Ethics of Research Related to Health Care in Developing Countries* (London: Nuffield Council on Bioethics).

Nuffield Council (2005), *The Ethics of Research Related to Health Care in Developing Countries: Follow-up Discussion Paper* (London: Nuffield Council on Bioethics).

Participants in the 2001 Conference on Ethical Aspects of Research in Developing Countries (2002), "Ethics: Fair benefits for research in developing countries," *Science* 298, 2133-4.

Pearson, H., (2004), "Summit calls for clear view of deposits in all biobanks," *Nature* 432, 426.

Singer, J.W., (2000), *Entitlement: The Paradoxes of Property* (New Haven, Conn.: Yale University Press), pp. 29-30, 83.

Schulz-Baldes, A., Vayena, E., Biller-Andorno, N. (2007), "Sharing benefits in international health research: research capacity building as an example of an indirect collective benefit," *EMBO Reports* 8:1, 8-13.

Swede, H., Stone, C.L. and Norwood, A.R. (2007), "National population-based biobanks for genetic research," *Genet Med* 9:3,141-9.

Sweden Law on Biobanks in Medical Care (2002, amended 2005) <www.sweden. gov.se/content/1/c6/02/31/26/f69e36Fd.pdf>, accessed 1 August 2007.

United Nations Educational, Scientific and Cultural Organizations (2003), *International Declaration on Human Genetic Data* <http://unesdoc.unesco.org/ images/0013/001312/131204e.pdf#page=27>, accessed 17 September 2007.

Winickoff, D.E. and Winickoff, R.N. (2003), "The charitable trust as a model for genomic biobanks," *NEJM* 349:1180-4.

World Health Organization (European Partnership on Patients' Rights and Citizens' Empowerment) (2003), *Genetic Databases, Assessing the Benefits and the Impact on Human Rights and Patient Rights*, Geneva 2003.

World Medical Association (2002), *The World Medical Association Declaration on Ethical Considerations Regarding Health Databases* (Ferney-Voltaire, France: World Medical Association).

PART II
CONSENSUS AND CONTROVERSIES AMONG INTERNATIONAL EXPERTS CONCERNING ISSUES RAISED BY GENETIC DATABASES

Chapter 4

Ethical Issues Regarding Research Biobanks: Aims, Methods, and Main Results of a Qualitative Study Among International and US Experts

Nikola Biller-Andorno, Andrea Boggio, Bernice Elger,
Agomoni Ganguli-Mitra, Alex Capron, Alex Mauron

1. Description of the study

This chapter describes the background and methodology of a collaborative study—initiated by the Department of Ethics, Trade, Human Rights and Health Law of the World Health Organization and the Institute of Biomedical Ethics of the University of Geneva with the subsequent collaboration of the Institute of Biomedical Ethics of the University of Zurich—on the conditions under which genetic databases can be established, kept and used in an ethically acceptable way.

1.1 Background

The technical possibilities for automated analysis of large DNA sample collections and the bio-informatic processing of the resulting data have developed dramatically during the past several years and are constantly being improved. Protecting the data available in such databases has consequently emerged as a highly complex ethical issue in the arena of health policy. The ethical issues become even more acute when genetic data are combined with information on individuals' health, lifestyle or genealogy. The political processes leading up to the establishment of these databases, as well as their particular legal and ethical arrangements, have stirred considerable controversy, highlighting the urgent need for an ethically grounded global regulatory framework. In spite of the value of directives developed by several countries and international organizations, few regulatory frameworks for genetic databases have been developed to date that are global in scope, yet developed with regional input, and that possess sufficient specificity to provide practical guidance. An important—and previously unmet—methodological challenge to developing such a global ethical framework lay in involving countries from different regions and stages of economic development. By engaging respondents from a wide range of countries with this intricate and controversial topic, the present project seeks to enrich the intercultural dialogue on the ethical aspects of genetic databases.

1.2 Objectives

As laid out in chapter 1, this empirical study aims to provide an account of the ethical arguments pertinent to genetic databases that is richer than what could be extracted from the existing literature and guidance documents, which are largely shaped by concerns of affluent Western societies. The study is explorative, probing the reasons for disagreement of those involved in the debate—scientists, biobankers, physicians, lawyers and ethicists from different parts of the world. We assumed that disagreement about ethical and legal issues concerning biobanks is influenced by socio-cultural aspects as well as by the particular interests and professional experience of different stakeholders. The objective of study was to capture the range of attitudes concerning the conditions under which genetic databases can be established, kept, and made use of in an ethically acceptable way. Our analysis of the respondents' different positions and the reasons they gave aims to deepen the discussion of existing barriers to a global framework for genetic databases and to guide further efforts in the search for resolution of current disagreements.

1.3 Methodology

The project is a social science study that utilizes a qualitative methodology principally based on semi-structured interviews with 87 respondents[1] worldwide. Interviewing is often referred to as "the favorite methodological tool of the qualitative researcher" (Denzin and Lincoln 1994, p. 363). And indeed, the wealth of information that interviews can produce, together with the flexibility of the semi-structured format, which allows researchers to probe unclear or particularly interesting responses, made this the appropriate methodological tool for our study.

Respondents were divided into two samples, an international sample (42 respondents) and a US sample (45 respondents). The international sample consists of experts selected from around the world and stratified according to continents and socio-economic differences; the US sample is limited to respondents working in the US at the time of the study, selected to include different ethical positions. Overall, respondents selected to participate in our study were chosen to reflect a wide range of opinion. Because of the small numbers, the respondents cannot be said to be representative for single countries, but the methodology justifies exploring differences between comparisons of the attitudes of some important subgroups. Almost two thirds of the respondents were from what could be called the socio-cultural background of "occidental" and relatively rich countries (the US, Canada and Western Europe). These regions are known to exert a predominant influence on the debate in the literature. It is therefore of interest to identify whether the attitudes of respondents from outside the highly influential North American and Western European cultural context differ from the mainstream debate in the literature on ethical and legal issues concerning biobanking. These "deviant" opinions might well enrich the discussion, bring out new aspects and improve understanding of

---

1 For simplicity's sake, all respondents, independently of their gender, are referred to by male pronouns.

why stakeholders disagree. In addition, we compared the attitudes of the 31 respondents who have used or stored samples with the attitudes of the remaining majority of interviewees whose experience lies in making policy recommendations and analyzing ethical and legal issues without direct, "practical" involvement with samples; this comparison was intended to test the hypothesis that differences in reasoning between subgroups defined by the theory-practice gap might be at the origin of certain disagreements about ethical issues concerning biobanks.

All interviews are based on four case scenarios that depicted the various circumstances and policy options involved in establishing and managing genetic databases. To insure consistency and to facilitate the analysis of responses, an interview protocol was developed that guided the interviewers through the process and structured the relationship with the respondents. The scenarios, policy questions and interview protocols were developed following an extensive review of the existing literature (especially documents that provide guidance on genetic databases) and through discussions with experts in the field (see below). The methodology was tested in six pilot interviews.

Data were collected through semi-structured interviews conducted on the phone and in person by members of the research team. Telephone interviews have been extensively used in empirical studies on health care and medical research (Spitzer et al. 1994; Emanuel et al. 1999; Wendler et al. 2002; Lanie et al. 2004; Kass et al. 2004). Combining telephone interviews and face-to-face interviews allowed us to efficiently reach a wider group of respondents located in different regions of the world.

The interviews were conducted by members of the research team.[2] The scenarios were distributed to respondents beforehand, as an attachment to the invitation letter, and respondents were asked to read the scenarios before the interview took place. The interview protocol was designed to allow the interviewer to pursue a respondent's reasoning through follow up questions while maintaining consistency among interviews. In fact, the diversity of opinions on biobanks as well as the different cultural and professional backgrounds of the respondents required us to be able to probe the respondents beyond the answers given initially. Each interview was taped; they lasted between 33 minutes and 120 minutes, the mean being 60 minutes. Respondents' personal identifiers were removed at the time of the interview and replaced by a number. During the entire analysis of the data, interview notes and tapes were referred to only by these numbers. (The tapes have been kept confidential and will be destroyed in due course.) The recorded interviews were transcribed.[3]

To allow the qualitative data to be analyzed, a system for coding responses was developed and used to code the content of each interview. This classification system was based on the existing arguments in support of the policies proposed in the scenarios; the coding categories were refined during the course of the data-analysis to better reflect the content of the responses. As a check on accuracy, a

---

2 B.E. conducted the interviews with respondents in the US sample. A.B. interviewed most of the respondents in the international sample, the majority in English and a few in French, Italian, and Spanish (the latter being conducted by a WHO intern, Sandra Realpe) in order to accommodate respondents who were not comfortable being interviewed in English.

3 The interviews were transcribed by members of the research team (B.E., A.B. and A.G.).

second member of the project team listened to each tape to validate the respondent's recorded answers and to review the correctness of the transcript and the coding for each interview.

Financial support for this project was provided by the Geneva International Academic Network (GIAN). Prior ethics approval for the study was granted by competent bodies at the University of Geneva and at the World Health Organization.

A closer look at the international sample   The respondents involved in the international sample were invited to participate in the study because of their professional background and their geographical location, a selection method known as purposive sampling. Purposive sampling is often used in qualitative studies to identify groups of people with specific characteristics or circumstances (Patton 2002; Dornan and Bundy 2004; Kenen et al. 2004; Hunt et al. 2001; Kumar and Gantley 1999). In purposive sampling,

Table 4.1   Characteristics of the international sample (n = 42)

Age	51 +/- years 11, minimum = 29, maximum = 77
Gender	Male: n = 30 (71%)
Affiliation(s)*	University/University Hospital: n = 23
	Government: n = 8
	NGO: n = 3
	Biobank/Database: n = 4 (private: n = 1)
	Independent Consultant: n = 4
	National Ethics Committee: n = 3
	International Organization: n = 1
Country of current employment**	Argentina, Austria, Brazil (n = 5), Colombia, Estonia, France (n = 2), Gambia, Hong Kong, India (n = 3), Israel, Japan (n = 3), Kenya (n = 2), Lebanon, Mexico, Netherlands (n = 2), New Zealand (n = 2), Nigeria, Spain, Sudan, Sweden, Switzerland (n = 2), Taiwan (n = 3), Tonga, United Arab Emirates, UK (n = 2), USA
Human Development Index***	High (≥ 0.8): n = 18 countries
	Medium (0.8 <> 0.5): n = 6 countries
	Low (< 0.5): n = 3 countries
Major field of expertise*	Life Science/Genetics: n = 17
	Medicine: n = 18
	Bioethics: n = 7
	Law: n = 5
	Philosophy: n = 4
	Others (Social Science/Political Science/Engineering): n = 5
Nature of work in this topic*	Analyzing Ethical and Legal Issues: n = 29,
	Making Recommendations/Drafting Guidelines: n = 16,
	Handling/Using Samples/Genetic Data: n = 18

* Some respondents indicated more than one affiliation, field of expertise or nature of work
** Nationality matched country of current employment except for 5 respondents
*** Figures taken from UNDP report [http://hdr.undp.org/statisti]

researchers choose subjects to participate in the study based on identified variables under consideration. In the present study, respondents were selected from individuals who have been previously exposed to the debate concerning human genetic databases. Furthermore, because of the cross-cultural nature of the inquiry and in order to facilitate the identification of common patterns across national variations, invited respondents reflected a wide range of professional backgrounds and geographical areas. About 50% of those approached agreed to participate: 90 invitation letters were sent, and 42 respondents consented to participate and were interviewed between March 2004 and October 2005. About two-thirds of the interviews took place in person and one-third over the phone; a few interviews started in person and were finished over the phone.

A closer look at the US sample   The data for the US part of the study were collected through interviews with 45 ethicists and scientists working in the US at the time of the interview. The interviews took place between March and June 2005. Following the purposive sampling approach, the group of respondents was selected to include scientists who had been involved with biobank research and ethicists who had published in the field or had dealt with the questions surrounding biobanking as members of IRBs and expert panels, as well as some ethicists without specific expertise in the area of biobanking. Geneticists and scientists with experience with biobanking were identified from major publications and from speaker lists at conferences in the field, as well as using a snowball system, in which experts were asked to indicate others whom they regard as experts in the area. The ethicists in the sample were working at major US bioethics centers with differing orientations. The sample was chosen to represent a variety of attitudes among ethicists and scientists, for example, to capture well known differences among ethical "schools," religious or philosophical orientations, and different professional backgrounds, in particular law, philosophy, medicine, social science and genetics. More experts were identified in this way than could be included in the sample, and the final selection was made on a first-mentioned, first-contacted basis, taking into account geographical convenience for carrying out the interviews.

The decision to include in the US sample ethicists who are not particular experts on biobanks was based on a desire to have respondents who were skilled in ethical reasoning but whose views would not be influenced by intensive study of any people involved in such discussions. On the one hand, if it is felt that good policymaking is dependent on knowing more details about certain controversial issues, it would be important to see what information needs to be more systematically given to the public and participants in biobank studies. On the other hand, biobank experts may fail to appreciate certain ethical issues simply because they are too focused on the specificities of the debates about biobanking and too influenced by other biobanking experts' policy propositions, meaning that they could benefit from taking account of overlooked views expressed by ethicists who have not extensively studied this area.

About 90% of experts and ethicists approached agreed to participate. About two-thirds of the interviews took place in person and one-third over the phone; a few interviews started in person and were finished over the phone.

Table 4.2　　Characteristics of the US sample (n = 45)

Age	54 +/- 11 years, minimum = 33, maximum = 85
Gender	Male: n = 28 (62%)
Race	White: n = 34
	Black: n = 4
	Asian: n = 2
Nationality*	US (n = 40, including 1 born in Nigeria & 1 dual citizen, Ireland), UK, Europe other than UK, India, Israel, Jamaica
Affiliations (current employer)	Hospital: n = 1
	NGO: n = 1
	Government: n = 11
	University: n = 11
	Medical school: n = 21
Geneticists/scientists	Working in genetics: n = 8
	Working in genetics and ethics: n = 2
	Working only in bioethics: n = 4
	Not practically involved with samples: n = 13
Major field of expertise**	Medicine: n = 9
	Life sciences including genetics: n = 9
	Law: n = 5
	Philosophy: n = 11
	Social science: n = 3
	Bioethics: n = 12
	Other humanities: n = 1
Country where highest degree was earned	US: n = 40
	UK: n = 2
	India: n = 1
	Germany: n = 1
	Israel: n = 1
Nature of work regarding ethical or practical issues of biobanking	Not particularly involved: n = 11
	Collect, use or store samples: n = 13***
	Analyze ethical issues: n = 27
	Make recommendations concerning biobanks: n = 24

*The country of current employment for all respondents in the US
**Five participants indicated more than one major field
***One ethicist-physician had worked with samples, and two geneticists said they were never significantly involved in collecting or storing samples

Abbreviations Used to Describe Respondents

The following abbreviations are used to categorize the respondents in five domains following the respondent numbers (which are consecutive from #01, except twelve numbers that were omitted due to the separate process of gathering the international and US samples):

1. Sample: IS = international sample, US = United States sample.
2. Region: For the international sample, the region of highest degree is stated first followed by the region of nationality and place of work; in the two cases where nationality is different from the place of work, this is indicated in brackets; when all three are the same, the region is only indicated once. In the US sample, this domain is left blank for all those who are North Americans; for those who recently arrived from another continent, the region is indicated after "US."
3. Major fields (may be more than one).
4. Place of employment: All participants labeled as "government" are from nationally funded research hospitals or institutions.
5. Type of involvement with biobanking issues (may be more than one):
   R= make **r**ecommendations
   S = use and/or store **s**amples
   A = **a**nalyze ethical and legal issues
   O = **o**ther involvement
   N = respondents with **n**o direct involvement with biobank issues

2. Outline of main results

In summary, the interviews show that the ethical issues concerning biobank research are indeed controversial. The responses indicate areas of relatively strong agreement on the following issues. Areas of disagreement and persisting controversy are explored in more depth in Chapter 5.

- While respondents generally agreed that irreversible anonymization of samples at the time of collection and storage is inadvisable, their responses showed that the terms used in current practices are often interpreted differently in various contexts, which results in apparent disagreements regarding "anonymization." Thus, standardized terminology may be needed if such frameworks are to be smoothly implemented.
- A degree of control should be exercised over the circulation of genetic samples, for instance through enforceable Material Transfer Agreements or other forms of monitoring, though the particular control that any stakeholder is allowed to have and the inclusion of particular groups (e.g., participants, communities, nations, researchers, sponsors) among the controlling stakeholders remain contested.

- The distinction between publicly and privately funded projects is very significant, although the precise consequences of this distinction remain quite controversial. Factors unrelated to biobanking—such as the economic situation of a respondent's country or its historical circumstances—often play a very important role in the views of respondents about the public and private sectors.
- Placing research results in the public domain within a reasonable time-frame is considered essential, although many respondents recognize that having some control over findings provides an incentive for research. The conflict between research freedom and population benefit is a familiar problem in ethics that takes on a particular aspect in debates about genetic data. For example, for someone to benefit from findings about a genetic propensity towards a certain illness may depend on interpreting risk probabilities, leading respondents to express differing views on who should be responsible for deciding what will be publicly revealed.
- A substantial number of respondents have reservations about "blanket consent" in its radical form (i.e., an agreement to allow any and all future research on a biological sample and/or personal information provided to a database). Such reservations suggest that biobanks may face a major stumbling block since a main objective in establishing such databases is the opportunity they provide researchers to access a large collection of material for studies that were not yet conceptualized at the time the materials were deposited in the biobank.
- Respondents disagree about what kind of "re-consenting" procedure is both ethical and practical for the use of samples in studies that were not foreseen at the time of initial consent.
- Although several benefit-sharing schemes are considered acceptable, direct payments to participating individuals is unadvisable. Consensus was lacking, however, regarding which forms of benefit-sharing are appropriate.
- The right to withdraw from a research project is a central feature of research ethics generally, but great controversy surrounds the question whether the sources of samples in genetic databases should always be able to withdraw their samples.
- The question of patrimonial rights in human genomic material is much contested in theory and in practice, as can be seen in the diverse responses about ownership of samples and the role of the database as custodian or as owner of samples and information. Views on this subject diverge in interesting ways between the industrial world and developing countries, in particular those with a colonial past.
- The operation of a genetic database can lead to incidental medical findings that are potentially relevant to the individuals who contributed to the database. There is little agreement about the duties that arise from these findings or about the proper roles for professionals when such information is conveyed to sample sources who may not have expected to receive it.
- Respondents have varying degrees of understanding regarding issues that do not directly touch upon their work context, such as collective consent or benefit-sharing. Responses on these topics may be rather one-sided since those who had not been much exposed to these issues did not have firmly held opinions.

All these issues call for more detailed analysis. In addition, it appears that some of the controversies about genetic databases stem from implicit disagreements about the applicable ethical framework: Are genetic databases essentially research infrastructures, or should they have a role in public health? The first view, dominant in the current literature, frames the issues that arise as problems of research ethics (i.e., subjects' autonomy, benefit-sharing, feedback to participants, and the like). The alternative view leads to an altogether different set of questions concerning what sort of public good is served by genetic databases, with more emphasis on issues of public accountability.

3. Scenarios and interview questions

The following four scenarios were used with both groups of respondents to probe their views about current controversies regarding ethically acceptable practices in biobanking. The scenarios were also translated into Spanish and French.

Scenario A

An international group of researchers has decided to establish and operate a repository in which DNA extracted from biological specimens will be stored along with related information. The purpose of the repository is to enable research on the association of certain genes with an elevated risk for developing colorectal polyps.

The research is funded by the health departments of three different countries, one of which acts as the home for the repository. The project is guided by a Steering Committee consisting of the principal investigator from each of the three countries, a representative of the international association for colon cancer patients, and a member of the Medical Research Council of the home country. An independent Ethics Review Board provides ethical guidance.

Local physicians will collect 3,000 biological samples from individuals diagnosed with colorectal polyps and from their blood relatives. The physicians will also fill out a health and lifestyle information sheet about every participant and send it along with the sample to the repository. This information will be periodically updated and any deaths will be reported. Informed consent will be obtained from all participants for their biological samples and associated information to be submitted to the repository. Biological samples will be stored at the repository in the form of extracted DNA.

A number of policies to guide the operations of the repository are now being discussed.

1. The Steering Committee has requested the advice of the Ethics Review Board on the following proposed policy regarding the protection of the participants' privacy:

   (a) Before shipping samples and associated information to the central repository, the physicians who collect the samples shall code the samples and associated information by assigning a number to each participant (1st-Series Code). The physician shall keep a record of the personal identities and their corresponding codes but shall not provide this information to the repository.

(b) Once each sample and the accompanying data have been received, a scientist not otherwise involved in the operation of the repository shall remove the existing code and assign a new number to each sample and associated information (2nd-Series Code). The scientist shall keep a list of the 1st-series codes and the corresponding 2nd-series codes in a secure location outside the repository.

(c) As additional information about existing participants is received, the independent scientist will in each case replace the 1st series code with the appropriate 2nd series code.

(d) Samples and data shall be identified only by their 2nd-series codes when provided to researchers.

→ *Do you think that the coding system described adequately protects the interests of the participants?*

→ *Is such a coding system necessary?*

→ *Could the coding process be left entirely in the hands of the repository?*

→ *As an alternative, a member of the Steering Committee suggests that the repository should irreversibly anonymize the samples and associated information. Do you favor this suggestion?*

---

2. The Steering Committee is deciding whether and, if so, to what extent participants can withdraw samples and associated information from the repository. Some members think allowing withdrawal would unduly burden researchers, while others argue withdrawal should be allowed up to the point where findings have been submitted for publication.

The Chairman of the Committee identifies four alternatives that committee members may choose from:

(a) Once samples and associated information are submitted to the repository, participants may not exercise their right of withdrawal regarding this material. The withdrawal would only be effective regarding the transfer of further material or new information.

(b) When a participant requests the withdrawal of samples and associated information, they will be irreversibly anonymized and continue to be used.

(c) When participants withdraw consent, their samples and associated information will be removed from the biobank and destroyed.

(d) When participants withdraw consent, their samples and associated information will be removed from the biobank and destroyed. Any researcher to whom they have been provided will be obliged to destroy the samples and to remove the samples and associated information from any report of research that has not yet been submitted for publication.

→ *Which of these alternatives would be the best policy? And why?*
→ *Would you rule out any of the alternatives?*
→ *How would you rank the acceptable alternatives from most preferable to least preferable?*
→ *If a participant dies, should the participant's legal representative be able to exercise his or her right of withdrawal?*
If not, what would be an appropriate policy for the right of withdrawal of data or samples after a participant's death?

---

3. The repository has already received several requests from investigators to use its materials in research not related to colorectal cancer. The representative of the Medical Research Council on the Steering Committee suggests that the participants would probably find such requests acceptable since they obviously support biomedical research. She therefore suggests that the Steering Committee adopts the following policy, which would construe the existing consent— *"I agree to Doctor ____ submitting my biological sample and associated information to the repository"* —as allowing the use of the samples and associated information in all projects approved by the Ethics Review Board:

All samples and associated information will be made available for approved biomedical research projects not only on colon cancer but also on other diseases.

→ *Do you think the policy proposal is acceptable—that is to construe the informed consent so as to permit research on other diseases besides colon cancer?*

---

4. Another member of the Steering Committee agrees that the present consent form is vague but she suggests that, rather than adopting the proposed policy, the original consent form should be modified. She notes that there are at least four alternatives to be chosen from:

(a) For each new study unrelated to colon cancer, participants will be recontacted and asked for their explicit consent
(b) For each new study unrelated to colon cancer, participants will be informed and offered a chance to opt out
(c) Participants will be explicitly asked initially whether they agree to their samples and associated information being used for research not limited to colon cancer.
(d) The consent form should be modified to allow the participants to choose from the options mentioned above.

→ *If NO to question on A.3: Since you did not favor the policy I just asked you about, would you favor instead any of the alternatives? Why?*
→ *If YES to question on A.3: I know you favor the policy that we have just discussed. What do you think of the four alternatives favored by the Committee*

member? Why?
→ *Would you rule out any of the alternatives?*
→ *How would you rank the acceptable alternatives from most preferable to least preferable?*

---

5. One of the investigators sitting on the Steering Committee proposes the following policy:

When specific gene mutations linked to an elevated risk of developing colon cancer are discovered, the repository will notify the physician who provided the coded sample and request that the participant be informed of the gene mutation and of its implications for his/her health.

→ *Do you think that this policy should be adopted?*
→ *Should a provision be added to the original consent form offering participants the option of being contacted through their physicians if they are found to have a gene mutation linked to an elevated risk of developing colon cancer (if they decline, no information would be sent either to them or their physicians)?*

If YES:

• *The policy provides that feedback is required only when a gene mutation linked to an elevated risk of developing colon cancer is discovered. Should feedback be limited to findings that are known to improve clinical management?*
• *Should findings that are known to improve the clinical management of diseases other than colon cancer be communicated?*
• *What about communicating information not related to clinical management but that might be relevant to reproductive choices?*
• *What about communicating information not related to clinical management but that might be relevant to planning one's life course (investment planning, work and family activities)?*
• *What about communicating information not related to clinical management but that might be relevant to paternity?*
• *Does the repository have an obligation to ensure that the physician notifies the participant of the findings?*

---

6. The Steering Committee proposes to include the following provisions in the Material Transfer Agreement (i.e., the contract setting the terms for giving an investigator access to biological specimens and the associated information):

(a) Investigators must not transfer the DNA and the associated information to persons not named in the Material Transfer Agreement.
(b) Investigators are under the obligation to share all research findings and the data produced for each sample with the repository.

→ *Is proposed policy (a) desirable—that is that investigators are prohibited from providing the DNA and the related personal data to others?*
→ *Is the proposed policy needed?*

If YES: Would such a policy be practical or would it be too difficult to enforce?

→ *Under policy (b), investigators must share research findings and data with repository. Do you think this is a desirable part of the agreement?*

If YES: Is the proposed policy a required part of the agreement?

---

7. A member of the Steering Committee has proposed the following policy:

   (a) *Ownership of samples* The repository is the custodian of the samples and associated information on behalf of the participants.
   (b) *Ownership of the data* The data generated by the research will be treated as a public good and all investigators must agree to put their findings and supporting data in the public domain on a regular basis and without undue delay.

The representative of the colon cancer patients proposes the addition of a provision under which investigators agree not to exercise any rights they may have to patent a gene sequence.

→ *The proposed policy provides that the repository is the custodian of the samples while each participant is the owner of his/her sample. Is this preferable to the repository being the owner of the samples once they have been submitted?*
→ *In other research projects, investigators are free to choose whether and when to publish data. Would you support the requirement that all data generated by the study be put in the public domain without undue delay (for example, submitting data to a publicly accessible genetic databank)?*
→ *What do you think of the proposed provision under which investigators agree not to exercise any rights they may have to patent a gene sequence?*

---

8. The Steering Committee is considering the following policy on the disposal of the samples:

Once the project has completed its study on colorectal cancer, the repository will be closed and the samples and associated data destroyed.

As an alternative, the representative of the Medical Research Council on the Committee proposes that, as long as her country is willing to maintain the repository, the samples and associated data should not be destroyed even after the colorectal cancer study has ended.

→ *Do you favor either of these proposed policies? Why?*

Scenario B

DNA Foundation is a non-profit organization established by the government of Country A for the purpose of permitting research based on the specific genetic characteristics of its population.

DNA Foundation has collected 200,000 samples from adults living in Country A and stored them in a central repository (located in Country A). The Department of Health has submitted the following policy proposal to the Foundation's Board:

The samples and the associated data cannot be transferred out of Country A.

→ *Do you think territorial limitations on the circulation of biological samples are ethically acceptable?*
→ *What about territorial limitations on the associated personal data?*
→ *Some people argue that, since "the human genome is part of the common heritage of humanity," such limitations are unacceptable. What do you think about this argument?*
→ *Some people argue that, in order to build up capacity, a country could exploit the existence of a genetic repository to encourage researchers to come to their country to do their work. What do you think about this argument?*
→ *Some people fear that confidentiality cannot be protected once samples leave the country and that therefore territorial limitations are acceptable. What do you think about this argument?*
→ *Assuming the participants were not paid, is it acceptable for the repository to ask researchers for a fee that is greater than what is needed to cover the costs?*
→ *If a fee beyond reimbursement is paid, should some of the additional income be shared with the participants who provide the samples and personal data?*

Scenario C

Biotech Incorporated proposes to collect 2,000 biological samples from the members of an indigenous population in the country where Biotech Incorporated is based. Pharma A, a publicly traded pharmaceutical company, finances Biotech Incorporated's project because it is interested in developing a genetic test to detect polymorphisms that are linked to adverse reactions to its most frequently prescribed drugs. These polymorphisms are known to occur more frequently, though not exclusively, in the studied population. Biotech Incorporated will be the owner of all intellectual property rights arising out of its research.

1. In its negotiations with the indigenous group, which is represented by a Governing Council, Biotech Incorporated has acknowledged its obligation to share the benefits of its research with the group. Biotech Incorporated is proposing various forms of benefit sharing to the Governing Council:

Option A: Making any genetic tests resulting from the research available for free to the indigenous group for ten years.

Option B: Making an annual donation for a period of ten years to the hospital that provides health care to the indigenous group, of a sum equivalent to 3% of the revenues generated by any intellectual property rights resulting from the research.

Option C: Donating several pieces of durable medical equipment to the hospital that provides health care to the indigenous group.

The Governing Council determines that none of the proposed forms of benefit-sharing are adequate, and that an agreement can only be reached if the Governing Council owns all intellectual property rights arising out of the research (*Option D*).

→ *Please rank the options of benefit-sharing from the most acceptable to the least acceptable:*

    a. Free genetic tests
    b. Sharing a percentage of profits
    c. Donation of medical equipment
    d. Ownership of IP rights

→ *Do you think any of these alternatives should be ruled out?*
→ *Are your choices based on considerations of practicality or on principle or both?*
→ *Some people would say that a number of these options amount to a fixed compensation for participation rather than to benefit-sharing. Is the distinction between receiving a fixed compensation and benefit-sharing morally relevant in the context of obtaining human biological samples?*
→ *Do you think any of the benefit-sharing options amount to a fixed compensation for participation?*

    a. Free genetic tests
    b. Sharing a percentage of profits
    c. Donation of medical equipment
    d. Ownership of IP rights

---

2. The negotiations become difficult and eventually stop. An employee from Biotech Incorporated, who is a member of the indigenous group, believes that the Governing Council is behaving arbitrarily and is out of touch with the beliefs of the group. She suggests to the head of the company that she could approach individual members of the group and offer them a sum of US$800 (roughly equal to four weeks of the average salary for members of the group), in exchange for their participation in the research. The offer would be contingent on the participant renouncing all intellectual property claims.

→ *Do you think Biotech Incorporated must first obtain permission from the Governing Council or may it approach individual members of the group directly without prior collective permission?*

If NO (collective permission not required): Is prior collective permission ever required?

If YES (collective permission required):

- *Under what circumstances is prior collective permission required?*
- *Each individual's characteristics allow him/her to be placed in different groups. Do you think prior collective permission would apply to all such groups or only to groups with specific characteristics?*
- *I will read you a list of characteristics that could require prior collective permission. Among the following, which one or ones do you feel is or are relevant, if any?*

  a. *The members of a group that has chosen to have a formal structure, with leaders.*
  b. *Traditionally, the group take collective decisions on issues that affect the whole group*
  c. *The group is economically disadvantaged*
  d. *The group is identifiable and the research results may be thought to apply to the group generally*
  e. *The group is ethnically distinctive*

Scenario D

Physicians at University Hospital routinely obtain informed consent from patients to store biological materials taken for future *diagnostic testing*. A law has recently been enacted requiring informed consent for the storage of biological samples for *research purposes*.

1. The Ethics Review Committee of University Hospital is choosing among three alternative additions to the standard consent form:

   (a) I agree to my stored specimen being used for medical research.
   (b) I agree to my stored specimen being used for medical research provided I have received advance notice and I have not opted out of the study.
   (c) I agree to my stored specimen being used for medical research provided I have consented to each use.
   → Do you think any of the alternatives are needed to comply with the new law on consent?
   → Should any of these alternatives be ruled out?
   → *Would the requirement of consent for research use apply if the specimens were irreversibly anonymized and unlinked to the patient?*

2. One of the members of the committee argues that it would be unduly burdensome to apply the consent requirement to samples stored at University Hospital before the law was enacted, and proposes the following provision be added to the policies on approving biomedical research:

> Where a research project involves linking a stored specimen to a patient who has not provided the necessary consent for the research, an investigator should recontact the patient to meet the consent requirements. If, however, a patient cannot be recontacted through reasonable efforts, the investigator may apply to the Ethics Review Board to use the specimen. The investigator shall show that reasonable efforts have been made and that the aims of the research cannot be achieved without using such specimens.

→ *Would you approve the proposed policy requiring reasonable efforts to recontact patients?*

→ *Do you favor the policy providing that, if reasonable efforts fail, researchers may use the samples provided they show the research cannot be achieved without using such specimens?*

Bibliography

Denzin, N.K. and Lincoln, Y.S. (1994) *Handbook of Qualitative Research* (Thousand Oaks, CA: Sage Publications).

Dornan, T. and Bundy, C. (2004) "What can experience add to early medical education? Consensus survey," *Br. Med. J.* 329(7470): 834.

Emanuel, E.J. et al. (1999) "The Practice of Euthanasia and Physician-Assisted Suicide in the United States Adherence to Proposed Safeguards and Effects on Physicians," *J. Am. Med. Ass.* 280(6): 507-13.

Hunt, K. et al. (2001) "Lay constructions of a family history of heart disease: potential for misunderstandings in the clinical encounter?," *Lancet* 357(9263): 1168-71.

Kass, N.E. et al. (2004) "Medical privacy and the disclosure of personal medical information: the beliefs and experiences of those with genetic and other clinical conditions," *Am. J. Med. Genet.* 128A(3): 261-70.

Kenen, R. et al. (2004) "Healthy women from suspected hereditary breast and ovarian cancer families: the significant others in their lives," *Eur. J. Cancer Care* 13(2): 169-79.

Kumar, S. and Gantley, M. (1999) "Tensions between policy makers and general practitioners in implementing new genetics: grounded theory interview study," *Br. Med. J.* 319(7222): 1410-13.

Lanie, A.D. et al. (2004) "Exploring the public understanding of basic genetic concepts," *J. Genet. Couns.* 13(4): 305-20.

Patton, M.Q. (2002) *Qualitative Research and Evaluation Methods*. Third edition. (Newbury Park, CA: Sage Publications).

Spitzer, R.L. et al. (1994) "Utility of a new procedure for diagnosing mental disorders in primary care. The PRIME-MD 1000 study," *J. Am. Med. Ass.* 272(22): 1749-56.

Wendler, D. et al. (2002) "Views of potential subjects toward proposed regulations for clinical research with adults unable to consent," *Am. J. Psychiatry* 159(4): 585-91.

Chapter 5

Consent and Use of Samples

Bernice Elger

1. Introduction

In this chapter we report the attitudes of the respondents from our study concerning one of the most central and controversial ethical questions raised by research involving genetic databases:[1] informed consent to the use of samples and the use of information related to the samples. After a short summary of the broader ethical debate in the literature and the positions taken in important international guidelines on this issue, participants' responses will be presented and the reasons for agreement and disagreement will be analyzed. This chapter examines the questions regarding the use of samples that have been stored in a biobank established for research. It is followed by a chapter on informed consent to research on samples from previously existing collections stored for other, especially clinical, purposes and a chapter on collective consent.

1.1 Informed consent and genetic databases: the challenge

To understand why consent to biobank research is controversial, one should remember the central role informed consent has played in the history of research ethics. Experiments on concentration camp inmates in Nazi Germany, and the Tuskegee Syphilis Study in which US physicians left patients uninformed and untreated to study the course of the disease resulted in worldwide shock and inspired energetic efforts aimed at protecting research subjects through appropriate regulation. As a result, informed consent became the gold standard of research ethics (Kegley 2004). Furthermore, research participants must be informed about all planned experiments for their consent to be valid. However, in the context of biobank research, entailing as it does the large scale, long term collection and conservation of samples or data, it is usually difficult to anticipate future research questions. Following the classical doctrine of informed consent, any consent to future research projects that are not clearly described, is by definition invalid because it is not informed. International guidelines on biobanks lack consensus as regards the importance and relevance of informed consent in this traditional sense.

1.2 Informed consent and genetic databases: published recommendations

In addition to Chapter 2 which has concentrated on international guidelines, we include here also national and professional statements, especially if they contribute interesting aspects to the discussion.

The extreme positions   The two extreme positions defended in the literature are "blanket consent" and new informed consent to each new project. The first approach has been proposed in a WHO publication:

> A blanket informed consent that would allow use of a sample for genetic research in general, including future as yet unspecified projects, appears to be the most efficient and economical approach, avoiding costly recontact before each new research project. The consent should specify that family members may request access to a sample to learn their own genetic status but not that of the donor (WHO 1998, p. 13).

The authors of this document explain further that "attempts should be made to inform families, at regular intervals, of new developments in testing and treatment. Donors should inform DNA banks of current addresses for follow-up" (WHO 1998, table 10, p. 13). In addition, two professional organizations have proposed a broad form of consent. The Human Genome Organisation (HUGO) stipulates that informed consent "may include … in some cases, blanket consent" (HUGO 2002, Art. 4(a)) and the European Society of Human Genetics favors "consent for a broader use":

> As it is difficult to foresee all the potential research applications that a collection may be used for, individuals may be asked to consent for a broader use. In that case, there is no need to recontact individuals although the subjects should be able to communicate should they wish to withdraw (ESHG p. S9).

At the other end of the spectrum of opinions, we find the "Genetic Privacy Act (GPA)," proposed by G. Annas et al. (Annas et al. 1995a and 1995b). According to this proposal, if tissue samples are identifiable, the person who provided the samples must give new consent in writing to all proposed use(s) of the sample once he or she has been informed about the details of these future uses. Although few others go so far as to accept the foundations of the "Genetic Privacy Act," i.e. the assumption that DNA samples are the property of the person with whom they originate (the "sample source"), many share its opposition to blanket consent. An example is the report from the American Society of Human Genetics (ASHG 1996). The ASHG encourages obtaining informed consent for all studies involving identified DNA samples, including prospective as well as all retrospective studies, "except if a Yes waiver is granted" (ASHG 1996, Table 1). The report explicitly states:

> It is inappropriate to ask a subject to grant blanket consent for all future unspecified genetic research projects … if the samples are identifiable in those subsequent studies (ASHG 1996, p. 471).

Intermediate positions   It is not surprising that the idea of new informed consent to each new project involving samples from a biobank has been defended in the legal community (Annas et al. 1995b). By contrast, researchers and others practically involved in biobanking claim that such a requirement is not only costly, but hampers research to the point that important studies are thwarted on account of the low percentage of research participants that can be reached after several years for new consent. Since "blanket consent" carries the connotation of abuse—of credulous research participants who provide samples without limits because they are not in the position to foresee future risks—numerous propositions have been made to find ethically acceptable solutions between the two extremes (Elger and Mauron 2003).

The position which seems to have gained the widest acceptance among North American commissioners is the "multilayered consent" proposed by the NBAC (NBAC 1999), the Tri-Council Policy Statement (Medical Research Council of Canada 1998, Art. 10) and the RMGA (RMGA 2000). The research subject or the patient is given the possibility to give or refuse consent to a large number of options.

> Suggested methods of handling secondary use of genetic material or research data include a comprehensive consent form, which allows the research subject to choose from a number of options…, or a more limited consent form, which specifies arrangements to maintain contact with the subject regarding future uses. Either method must be clearly explained during the free and informed consent process (Medical Research Council of Canada 1998, Art. 8.6.(p. 8.7.).

The options include permitting use of their samples only after irreversible anonymization,[2] permitting coded or identified use for one defined study, or for any study relating to the condition for which the sample was originally collected. These options are further qualified with or without permission to be re-contacted for other studies. Study participants have the choice to permit the use of their samples for research in some areas and to exclude others, for example research about addictive behaviour, from a detailed list. In contrast to most other guidelines, the NBAC (NBAC 1999) includes blanket consent for all other research as a further option and seems to suggest that this form of "consent" could become more acceptable if it is not offered as the only option, but as one among others.

Another intermediate strategy has been to permit some form of "semi-blanket" consent if the new research is in the same overall domain (CCNE 1995) or about the same medical condition (Medical Research Council of Canada 1998, Art. 10.3, p. 10.4.).

A third intermediate strategy, a form of "opt out" or "presumed consent" policy, can be found in guidelines from the Human Genome Organisation (HUGO 1998). According to this statement, "[r]esearch samples obtained with consent and stored may be used for other research if: there is general notification of such a policy, the participant has not objected, and the sample to be used by the researcher has been coded or anonymized." As noted by Deschênes et al. (Deschênes et al. 2001, p. 225), "[t]his procedure offers a theoretical right of refusal for the participant. It requires the creation of a mechanism by which the refusal of the participant could

---

2 For the terminology and ethical issues related to anonymization see Chapter 5.

be registered and observed." In other words, the HUGO ethics committee allows for a solution akin to blanket consent if participants have been notified and have the opportunity to object to further use of coded samples.

Although there is no agreement in the literature and among guidelines on consent to research involving biobanks, all intermediate solutions have in common that they depart to some extent from the classical doctrine of informed consent.[3]

1.3 Which type of consent is adequate for studies involving a typical long-term research biobank (Scenario A.3 and A.4)?

The issue of subsequent uses of samples for purposes not stated in the original informed consent form was touched upon in two scenarios (A.3/A.4 and D), giving two different contexts for reflection. This chapter is limited to the first (scenario A). The vignette describes a publicly funded repository under the responsibility of an international group of researchers. This biobank has been established to study the association of certain genes with an elevated risk for developing colorectal polyps. Participants' samples and information about their disease and lifestyle are obtained from physicians in three countries and stored and used after having been double coded.

We were interested in knowing respondents' attitudes towards different forms of consent. In scenario A.3, the hypothetical steering committee of the international biobank proposed a form of "construed consent", a term henceforth used to mean consent to sample collection construed as authorizing any future research. In scenario A.4 interviewees were confronted with four other options for consent for the same biobank: (a) new informed consent for each new study, (b) a form of presumed consent with included information on any further study and the possibility to opt out for further studies, (c) general consent for future studies and (d) layered consent offering the choice for study participants between options (a) to (c).

2. Results

2.1. Construed consent

In scenario A.3, it was assumed that physicians had obtained samples and information from patients and their families using the following vague consent: "I agree to doctor _____ submitting my biological sample and associated information to the repository." Respondents were asked whether it is acceptable to interpret this consent broadly as allowing the use of the samples and associated information in all future projects approved by the Ethics Review Board of the repository. The vignette stated that this broad interpretation of consent is based on the assumption that the sample donors probably find future research acceptable since they obviously support biomedical research. The policy proposed in scenario A.3 is that "all samples and associated

---

3 See also Chapter 3 for the intermediate solutions proposed in other recommendations (e.g. from the Council of Europe 2006 and the WHO European Partnership on Patient's Rights and Citizens' Empowerment 2003).

information will be made available for approved biomedical research projects not only on colon cancer but also on other diseases."

Most respondents agreed that the policy is not acceptable. One scientist admitted not being able to answer this question because he is "torn between being a scientist and being on the ethical side" (#61 IS, Africa and Europe/Africa, life sciences, university, R/U/A/O).[4] This answer is characteristic for the opposition of two values with which all respondents struggle: usefulness of the samples for research and the ethical problem caused by the lack of informed consent. Indeed, the disagreement between respondents is explained to a large extent by the fact that for most respondents, autonomy rights, i.e. the right to consent to the use of samples, trump the interests of research in the case of this particular international biobank. Only a few, most of them interviewees involved actively in research on biological samples, argued that the policy in scenario is acceptable, although it should be noted that almost all of them limited the scope of the construed consent and wanted additional conditions to be fulfilled.

Arguments in favor of construed consent   First of all, as one participant from the US noted, one has to differentiate between two questions: can you do research or can you construe the consent: "I think you can do research without consent in such a case [but] you cannot construe the consent." Although construed consent is unacceptable, it is a different question to ask whether one could use the samples without consent under certain circumstances (#59 US, medicine, bioethics, government, R/A/S).

Indeed, only one interviewee (from Europe) defended construed consent as such. This respondent would accept construed consent in the case that the consent form was truly vague and did not suggest a restriction to colon cancer:

> If the informed consent form does not mention any form of research, I would say 'yes'. If the form mentions only colon cancer, then I would not agree (#17 IS, Europe, law, university R/A).

Most other respondents argued rather in favor of the use of the samples than in favor of allowing construed consent. Most of them discussed conditions that would make the extended use of samples in scenario A.3 acceptable.

> It would depend upon the importance of the research, it would depend upon other reasons to think that the research might be considered controversial, I think it would depend

---

4 The following abbreviations are used to categorize the respondents (see Chapter 4): IS = international sample, US = United States sample (note: all are North Americans except some who recently arrived from another continent which is indicated after US). For the international sample: The region of highest degree is stated first followed by the region of nationality and place of work (if both are the same as it is most frequently the case, in the two cases where nationality is different from the place of work, #68 and #72, this is indicated in brackets). If all three are the same, only one regional indication is given. In the US sample, all participants labeled as "government" are from federally funded research hospitals/institutions. The respondents' different degrees of involvement with biobank issues are noted: R= making recommendations, S = sample use and/or storage, A = analysing ethical and legal issues, O = other involvement, N = respondents with no direct involvement with biobank issues.

on whether sufficient steps were taken to ensure the confidentiality of the subjects and whether opportunities [exist] to inform them of the broader uses of the research (#59 US, medicine, bioethics, government, R/A/S).

For example, a philosopher and theologian indicated that utilitarian arguments could outweigh the requirement for informed consent if the burden caused by recontacting is high as well as the benefit expected from the research:

> I would accept it [the policy in scenario A-3] on utilitarian grounds if there is some possibility to produce some benefit ... If it is impossible to recontact them [research participants] because you don't have any address then the requirement cannot be realistically fulfilled and if there is a serious benefit from the research then that should prevail (#02 US, philosophy, catholic theology, university, N).

Similarly, respondents in favor of construed consent defended their position referring to the value of the samples.

> Genetic materials are very valuable resources and they are very expensive to collect. Once you have collected them, it is ... a waste to then just limit them to a narrowly defined set of studies (#81 IS, No. Am./Africa, medicine, bioethics, university, S).

According to a respondent from Asia,[5] in cases where biological material is scarce and of outstanding value, it is ethical to use it for other selected studies if the research is beneficial for the population to which research participants belong, even if consent had been limited to colon cancer.

> It has to be on a case to case basis. For example if you have a tribe going extinct, then the material becomes very precious so you cannot allow that to be used for [any] disease but if you know they suffer from a particular disease, for that disease alone you could use that (#42 IS, Asia and No. Am./Asia, medicine, bioethics, government, A).

Two participants argued that the broad use of samples is acceptable if research is limited to certain types of studies. Research on questions related to human biology in general should be permitted even in the case that consent was only provided for colon cancer.

> It is a problem for me because for me all biology is connected. And I don't see that you can isolate the underlying biology of colorectal cancer from others. I can expand it so that in a biomedical study you might use colorectal cancer as a control group as for something. So I would be inclined to want to review the application and agree to make this available if it meets a study set criterion that does not have to be looking for colon cancer but it has to do with understanding the biology of disease (#15 US, genetics, university, R/A/S).

The other participant gave the following example. He considers consent acceptable if the consent form listed the study of hormone metabolism genes "and then you go

---

5 A similar opinion was expressed by a respondent from an indigenous group who could approve of construed consent personally although he was not sure about attitudes of other members in his group.

off and study DNA damage repair genes—something not exactly what may have been intended in the original study but [what] is really similar, comparable." To compare with, this respondent provided an example for research questions for which construed consent would *not* be acceptable:

> But you could also imagine another study where you have these lymphocytes and you use them for human cloning studies; that is a fundamentally different kind of study. So you are using the samples for different studies that have not been intended initially but in the first case it is very close whereas in the second case it is fundamentally different where the ethical issues are completely different (#96 US, genetics, medicine, social sciences, university hospital, R/A/S).

Another condition that could make the extended use of samples acceptable in scenario A.3 for one respondent is adequate protection such as IRB approval or waivers (#10 US, humanities, law, university, R/A).

> [I]f the patient was informed that this is a colon cancer study and that the samples will be stored and so on, then the patients ... said yes to colon cancer and not specifically no to anything else ... So in principle, samples collected for specific studies can be used in our view for other studies if the ethical review board agrees to this even if the possibility of future studies was not mentioned in the consent (#32 IS, Europe, medicine, university, S/A).

Arguments against construed consent Most participants from the US and the international sample stated that it is not acceptable to construe the consent as described in scenario A.3, the main reason being that research participants had not received sufficient information to be able to imagine the future uses of their samples correctly.

> I think giving some open texture to the consent is fine but this has such an open texture that I think participants would not be able to imaginatively foresee what the researchers might have in mind (#16 US philosophy, bioethics, university, N).

Although many interviewees recognize how important it is for scientific purposes to use the samples for research on other diseases, they find this practice contrary to the right of individuals to decide about the use of their biological material, because in this case research participants consented under the mistaken assumption that their samples would only be used for research on colon polyps and cancer.

> I don't agree with this interpretation of the statement ... the project is tied to research on colon cancer, and so there is an agreement to only research on that disease (#13a IS, No. Am./Asia, philosophy, A/R).

> I think it is important to have that potential to go beyond the colon cancer but I also think that then the original informed consent form has to state that (#97 US, medicine, genetics, university hospital, R/A/S).

Others added that it is generally inappropriate to construe consent based on assumptions about research participants' wishes:

I think it is inappropriate to construe anything. Participants either consented to their use specific to colon cancer or they consented to the use of their samples in advanced biomedical research (#44 US, bioethics, social sciences, A).

They pointed out that it is wrong to assume that participants who agree to colon cancer research would automatically also agree to biomedical research in general:

[The policy] presupposes the core of participants participated in one specific type of research on colon cancer and that they then support biomedical research, and that's actually a huge assumption … it doesn't mean to say that they want to participate in anything else. That assumption should not be made (#14 IS, Oceania, political science, indigenous IP, university A/R).

Informed consent should be not expansively construed: just because I consent to 'a' I consent to 'b' there is no reason to presume precisely why I consented to 'a' that I would therefore consent to 'b'. I have my own reasons for consenting to 'a' and I have my own reason for not consenting to 'b'—it goes beyond any reasonable construal (#20 US, law, philosophy, university, R/A).

Whereas for most interviewees it would have been sufficient if the consent form had mentioned "future uses for other diseases," a substantial number of interviewees from the international sample opposed construed consent in scenario A.3 for even stronger reasons. They think that respect for research participants' autonomy implies that consent has always to be specific (see 2.2). Recontacting and re-consent to each distinct future use is considered the only equivalent of true informed consent:

Samples are collected for a specific purpose, and if this is changed, a new permission from the donor is required (#06 IS, So. Am, philosophy, bioethics, university, A).

It undercuts again prior informed consent. There is not such a thing such as a vague, broad consent that is truly informed (#46 IS, No. Am., law, social science, NGO, A).

Respondents criticized the policy not only because it is at odds with their understanding of respect for research participants' autonomy, but also because in their view it is against the patients' best interests to have their vague consent be interpreted as a "blanket approval" (#23 US, medicine, humanities, genetics, university, R/A/S). They seem to assume that it would be better for patients with colon polyps to stay in control about future uses of their biological material.

In contrast to the respondent who favored extended use of the samples with IRB approval, respondents opposed to construed consent said that IRB approval by itself is not sufficient to justify using samples without consent. The original consent form has to state that future uses with IRB approval are planned:

If it is clearly stated in the informed consent and the donor consents to it. It is ok as long as there is further specification that 'the samples will only be used in projects that have been approved by the ethical board' (#31 IS, Europe, medicine, genetics, university, biobank, S).

Finally it should be noted that numerous participants indicated spontaneously that it is regrettable, or even "foolish" (#36 US, medicine, genetics, government, R/A/SU) not to have mentioned a broader consent in the consent form.

> If they start off with a [narrow] consent they have to live with the consent they have granted. That's why I think they should expand the consent form right from the beginning and in a sense you can do that as long as you put in the consent form that nothing will be done that is not approved by the ethics committee of our project—that will certainly reassure people I would think (#63 US, bioethics, medicine, catholic theology, university, A).

2.2. New informed consent for each new study versus general consent, presumed consent or multilayered consent

While agreement was high as regards the rejection of construed consent, responses varied considerably when it came to choosing the best policy for consent to research involving the international biobank described in scenario A. The policy that was most frequently chosen and least frequently ruled out was general consent (option c). About half of the respondents from Europe and the US as well as from all remaining regions prefer this form of broad consent if research participants have been informed beforehand that their written consent covers further studies on other diseases. More than one fourth of the interviewees from outside the US and Europe, compared to less than 10% of respondents from the US and Europe, favored new informed consent for each future study (option a). Presumed consent (option b) was the option chosen least often, by only a handful of respondents. More respondents from the US and Europe (one third) than from other regions (approximately one fifth) consider multilayered consent the best option. Several interviewees were against all consent policies described in scenario A.4 and proposed somewhat modified options. Some respondents find general consent (option c) only acceptable if research participants were also offered the possibility to consent to research limited to colon cancer. A few participants indicated that multilayered consent has to include more detailed choices in terms of categories of research ranging from three categories, such as colon cancer research, research on other types of cancer, research on other diseases than cancer to a list of at least 10 different specified diseases.

Arguments in favor of general consent   This consent option was described in the following way in scenario A.4: "Participants will be explicitly asked initially whether they agree to their samples and associated information being used for research not limited to colon cancer."

Many respondents pointed out practical arguments in favor of general consent. This type of consent is the "simplest" (#59 US, medicine, bioethics, government, R/A/S), the "easiest" to put into practice among the consent options (#99 US, medicine, genetics, university hospital, R/S). These practical advantages represent a benefit for any biobank research in terms of costs and management. Respondents indicated also that general consent is beneficial for scientific progress because samples can be used for a maximum of different purposes. A more restrictive approach is considered an ethically inappropriate waste of resources.

> My perception is that there should be an explicit statement [for general consent, like (c)], that's the perfect approach ... I think it would be improper for the scientists ... if you restrict too much, then the science doesn't move and you are also spending a lot of money for storage. That's why (c) is the best option (#78 IS, Asia, life sciences, government, S/A).

However, according to many interviewees, the benefit is not limited to the biobank. General consent would not only "make researchers happy" (#26 US, medicine, social sciences, government, R/A), but is felt to be beneficial also to research participants. General consent is easier to understand and avoids confusion "because then you are not complicating the situation, you are being straightforward about the anticipated uses" (#44 US, bioethics, social sciences, A). Compared to other consent options, it offers research subjects the easiest way to express their choices "without having to worry again" (#59 US, medicine, bioethics, government, R/A/S), as they are not bothered later by recontacting or newsletters.

> Again you will find that most people who want to use it for one thing would be happy to use it for other things and so it allows people to exercise those choices and expresses choices for those individuals who have strong views about participating in research and want to see that happen; that allows them to exercise that choice without having to worry again (#59 US, medicine, bioethics, government, R/A/S).

Arguments in favor of general consent also mentioned respect for research participants' wishes. Respondents favoring general consent are convinced that most research participants prefer general consent to other more complex forms of consent. General consent "satisfies a large majority of people" (#26 US, medicine, social sciences, government, R/A).

> I would choose (c) [general consent]. Actually from my reading of the literature and the one study I participated in, the data indicate that most people really once they agree to have their samples used for research are willing to have it used for ... any research and I think it is more confusing and unduly burdensome to give people a whole menu of choices when they really don't care about it. They have already made a decision. They trust the researchers and as long as they know that it could be used for other conditions they have the option to either do it or not do it (#45 US, bioethics, philosophy, government, O).

It "seems to be what patients want, again not only in the US, [but] internationally [according to] studies in England, Sweden and the US" (#33 US, bioethics, medicine, government, A).

Respondents in favor of general consent are aware of the criticism that general consent might not be considered truly informed consent since detailed information about future research projects is lacking. However, respondents retort that even in this general form consent is informed, as long as research subjects have been explicitly asked at the beginning of the colon cancer study to agree to future studies on other diseases (#97 US, medicine, genetics, university hospital, R/A/S).

I consider it informed consent: you are agreeing to it [the sample] being used for other forms of biomedical research as approved by an ethics review board. Either you consent to this or you don't (#95 US, bioethics, philosophy, theology, university, N).

It is considered a valid form of consent because it is believed to fully respect research participants' autonomous choices, including the choice of not being contacted again. Respondents pointed out that it is in line with the ethical requirement of respect for research participants' autonomy to let people "decide that their samples can be used for a variety of quality research projects" (#26 US, medicine, social sciences, government, R/A).

From everything we know, patients can be fully respected without each time asking them or without each time asking them to opt out. And you don't need to give them a laundry list from which to choose (#33 US, bioethics, medicine, government, A).

Although the lack of information about future research projects is a recognized problem, general consent is preferred because it represents the most favorable balance between respecting research participants' autonomy and benefit to research.

I think that (c) is the best alternative, both in terms of being fair to the subject and at the same time allowing the best use of the samples in terms of advancing knowledge ... I think that (c) is preferable in terms of respecting people but not adding great expense and difficulty to the research (#36 US, medicine, genetics, government, R/A/SU).

However, several interviewees' acceptance of general consent is conditional on various safeguards, such as "substantial protections of privacy," a "sufficiently robust and transparent process" and "other procedural protections" such as IRB approval:

I know that there are some people who think that it is unreasonable to think that you can consent to any sort of research in the future without knowing what it is. I think that it is possible to consent to that in a setting where there are substantial protections of privacy and where there is oversight by an IRB and by an ethics committee and possibly some kind of community advisory board. So on balance I come down in the favor of (c) [general consent]. I think offering people the options in (d) [multilayered consent] is a great idea but I don't know that it is obligatory. ... I think given a sufficiently robust and transparent process, they can agree to other kinds of research given the other procedural protections (#92 US, law, medicine, genetics, university, R/A).

Others referred to the safeguards provided by the general legal framework in their country, which they apparently trust to prevent abuses: "[A]s long as it is legitimate biomedical research," research participants are perceived to be "comfortable with a vague general authorization" (#44 US, bioethics, social sciences, A).

A respondent from Asia indicated that it is not possible "to protect participants with informed consent" equally for *all* future projects (#13a IS, No. Am./Asia, philosophy, university, A/R).[6] He, as well as others argued that general consent should be "multi-purpose," but not extend to all types of diseases. For certain research fields re-consent should be obligatory.

---

6 Respondents #13a, #13b and #13c were interviewed together.

> [Our country] is going towards a multi-purpose consent ... at the beginning of the biobank project, it should be stated clearly in the informed consent what range and kinds of disease will be researched. If further on there is a need to expand the scope of research to a significant degree, then re-consent should be considered (#13c IS, No. Am./Asia, law, university, A/R).

Among the diseases or conditions that were considered unsuitable for general consent were violent behaviour, mental disease and HIV, i.e. conditions linked to social stigma.

> But now you want to study something about violence tendency, it's a whole different story. In my opinion it is necessary to get re-consent, and the question [on whether recontacting is required] should be decided by the IRB (#13b IS, No. Am./Asia, law, university, A/R).

Instead of positively including diseases in the general consent form, some prefer an exclusive approach, listing only those types of research to which the general consent does not apply.

> Basically I would not specify more [in the general consent form]. What I think about is to exclude certain diseases, for instance one may exclude HIV but I would not specify any research but exclude certain researches (#17 IS, Europe, law, university R/A).

The answers of numerous respondents show how the acceptance of general consent is linked to the question of trust. General consent will only be an acceptable option for research participants as well as ethicists if they have evidence that research involving genetic databases is carried out in a trustworthy way.

> Again they give their consent, they trust us. So we have to be at the level of their trust. We have to live up to the level of their trust. I think in ethics and in research a lot depends on that. If patients don't trust us, there will be no research anymore (#57 IS, No. Am./Europe, medicine, university, S).

Respondents in favor of general consent expressed trust that IRBs function as sufficient safeguards, as illustrated in the statement from a South American respondent:

> If the investigation is approved by an Ethics Committee, there is no need to put any restriction on [uses of samples in the informed consent form] (#79 IS, So. Am., medicine, university, S/A).

Finally, interviewees favoring general consent are convinced that possible disadvantages of this consent form exist, but are of minor importance. For example, general consent might imply that a certain percentage of research subjects decline participation in all research involving biobanks:

> [I]t is an empirical question to find out how many people would then say no [to general consent...] whereas the options [multilayered consent] would allow them at least to participate in some studies. ... I feel comfortable with a certain number of people declining to participate in studies. That is the purpose of the consent process to let them make this decision (#26 US, medicine, social sciences, government, R/A).

Arguments against general consent General consent was ruled out by few respondents (less than one seventh) from the US and Europe, but by about one fourth of the interviewees from outside these regions.

For several interviewees, this consent option was perceived to be "too general" (#41 US, bioethics, law, university hospital, R/A) and too close to blanket consent:

> [The uses] should be explicit and specified ... I am against a blanket consent (#21 IS, Europe/Asia, philosophy, university, A).

Both general and blanket consent are considered inappropriate because respect for autonomy implies that research participants should know how their samples will be used. The lack of information makes general consent ethically unacceptable, especially if future research might involve completely different diseases.

> I think (c) [general consent] is maybe too open because clearly I would not like the sample to be used for a completely different purpose .... But clearly when you study ApoE [Apolipoprotein E], you look at it as a determinant of the lipids, and suddenly it happens to be also a determinant of neurology disorders. ... genetic systems are so complex—they are not very specific—and so you can say that this system plays a role in different types of diseases. And so [this is] why I am in favor of asking the consent, saying explicitly what you are going to do (#57 IS, No. Am./Europe, medicine, university, S).

Hence, several respondents against general consent prefer to add information about the types of diseases that will be studied in the future:

> It [consent] must be more specific: 'do you want to research lung cancer? Other cancers?' playing as much as possible on specificity before [collection] (#53 IS, Europe, medicine, public health, Int. Org., R/A/S).

Respondents pointed out that general consent compromises respect for autonomy because the way the colon cancer biobank is designed makes it impossible for research participants to envision the consequences of their participation in research and to make an informed decision:

> I say no to (c) [general consent] because if they are still feeding in all that medical information and if this has no obvious end point then I just think it would be extremely difficult for a person being asked to sign to actually envision the implications of agreeing (#52 US, social science, bioethics, university, R/A).

Apart from the lack of respect for patient autonomy resulting from inappropriate information, respondents mentioned the risk for abuse:

> (C) [general consent] is just a blanket approval for the future, and it can be abused (#22 IS, Africa, science, self-employed, A; in favor of new informed consent for each new research project).

The risk of abuse is related to the vague definition of further research which could include future unethical studies.

> I guess it [general consent] is the most subject to abuse potentially because it depends on how you define research (#04 US, medicine, genetics, university hospital, R/S).

Respondents opposing general consent differ from respondents who favor it with respect to trust in procedural safeguards, such as IRBs. Distrust in ethics committees in his country is described in more detail by an interviewee from Africa.

> It [general consent] is very difficult because, in my opinion, it does not inform people, it is very vague. And this opens the door to rethinking and so forth, and in my opinion, this is not a positive thing. One must approach the people and explain to them from the beginning what is intended. Things that are too vague open the door to problems. ... In practical terms, these ethics committees are a mixture of genuine interest in ethics and [interest] in the political aspects of the issue. It thus becomes an instrument to exercise power and control. Therefore, one must be careful ... especially when there are members from the government, whose agenda is never explicit (#56 IS, Europe/Europe and Africa, medicine, life sciences, government, hospital, S/A/R).

Insufficient trust in IRBs is not only mentioned by interviewees from Africa, but also in the US. Informed consent is said to be important not only in principle, but is necessary to protect research participants from future harm.

> I don't think this [general consent] is a responsible way to seek subjects for experiments I think they ought to know what they are participating in and I think that in principle and I also think this for the protection [Even with IRB approval] because I don't trust IRBs. I think IRBs very often let things through because these same people are going to be evaluating their research protocols and I know IRBs. I have colleagues now who serve in my institution who get stacks that are 3 feet tall once a month and they are assigned maybe to review one particular one in depth but I don't think IRBs are sufficient safeguards (#09 US philosopher, catholic theology, university, N).

According to one respondent from South America, it is not the risk of abuse itself that makes general consent ethically unacceptable. Instead, the human rights violations of research subjects in the past have raised public awareness and elicited the strong perception that standards for subject protection need to be high. Hence, any form of general consent has the connotation of being too close to past unethical standards:

> It is like signing a blanket check. I am aware that from a practical point of view this [new informed consent for each use] is problematic, but, as of today, it must be like that. Perhaps, in the future, with new developments, when we will have forgotten what doctors and physicians did, one could give an almost blanket consent (#06 IS, So. Am, philosophy, bioethics, university, A).

For a respondent from Africa, exaggerated fears about risks of abuse in genetics make IRBs in his country refuse protocols using general consent. Hence, this respondent does not favor general consent because of the resulting practical problems to obtain IRB approval:

> Here [in Africa] a blanket approval is seen as problematic by ethics review boards. We try to avoid [them] to the extent [this is] possible ... because we noticed that a lot of time is

wasted in endless discussions, and then everything stops. [Ethics review boards] are a mix of researchers, lay members, who read Time Magazine and Newsweek and become aware that genes can be something bad, and thus pointless discussions arise. It is a heterogeneous committee where some members know something, and other members ignore everything about genetics. What happens here is that the ethical discussion has not progressed in parallel with the increase of scientific information in the field of genetics (#56 IS, Europe/ Europe and Africa, medicine, life sciences, government, hospital, S/A/R).

In contrast to respondents in favor of general consent, respondents opposed to it are convinced that participation in research on different diseases is not in line with research participants' wishes and motivations and could therefore amount to exploitation.

I rule out (c) and (d) [general consent and multilayered consent], because it undermines the perception of the integrity of the original research and the motivation for the original research .... It makes it look like we are after your material and we may do with it whatever we want. It seems to disrespect the original reasons why someone was motivated to join the original study. From the subject's perspective, unlike for the researchers' perspective there are huge differences between different diseases because they have a connection to one but not necessarily to the others (#55 US, philosophy, bioethics, university, N).

This respondent believes that research participants give samples because of their "connection" to a particular disease that has occurred in their family such as colorectal polyps and colon cancer as described in scenario A. Asking them for consent to use their samples for research on other diseases takes advantage of their original motivation and might put undue pressure on them, as explained by another respondent:

My worry about (c) is if it is written in a way that makes it sound like or makes people feel pressured to say yes it's OK in other words if they do want to participate in the original study—let's say they had a relative die of colorectal cancer and they really want to help that cause—they might feel pressured to say: oh sure it's OK to do anything else. So you really want to make sure there is an option. So (c) [general consent] is the distant last and (d) [multilayered consent] is the heavy favored (#16 US, philosophy, bioethics, university, N).

Finally, one respondent opposed general consent because of perceived negative consequences for the scientific value of the future research projects. Interestingly, this respondent had never used or stored samples.

With (c) you run a chance of having too restrictive a bank of samples you might not be able to do as thorough a study and you might exclude people who would be willing just do it for the colon cancer, so I would put (c) close to the bottom (#28 US, bioethics, medicine, government, N).

Arguments in favor of new informed consent to each new study   In scenario A.4, this consent option was described in the following way: "For each new study unrelated to colon cancer, participants will be recontacted and asked for their explicit consent."

Most respondents who favor this type of consent have not used or stored samples. Arguments mentioned in favor are first of all related to the respect for autonomy.

New informed consent to each new study is judged "the most respectful for the participant" (#1 US) because it is the only option to provide true informed choice. The answers of the interviewees reveal three main reasons for this preference. First, although the practical difficulties of this consent type are acknowledged, these respondents are convinced that information and samples belong to the patient and that patients' rights are more important than the difficulties for researchers:

> It [(a), i.e. new informed consent] is the only one that gives the patients the option to choose and that [assures that] the perspective of the person from whom the sample was taken is always taken into consideration, in relation with his/her interest, expectations, and with the possible development of the study. ... It [re-consent] is difficult because usually investigators do not have the time to do it, [i.e.] search for the patients, and also they are not willing to do it. But under this kind of circumstances, we think it is necessary because the information that is there belongs only to the patient (#86 IS, So. Am., medicine, university, A).

Second, new informed consent is judged to be the only way to respect participants' wishes and to obtain their enrollment in genetic studies because research subjects don't trust biobanks in the present climate of genetic discrimination and stigmatization:

> I would favor (a) [new informed consent]. Until trust is established and that will take a long time from now because now everyone is worried about stigma, it is worth the effort to recontact everybody to get permission (#55 US, philosophy, bioethics, university, N).

Third, respondents in favor of this consent type are convinced that it is practical:

> I think that when they start the research they have all the information regarding the participants, and I think that there is no reason why they should not be able to recontact the person (#22 IS, Africa, science, self-employed, A).

Interestingly, among respondents who find recontacting for new consent practical, only one used or stored samples. He described why, according to his experience, re-consent is feasible in a particular research setting of a brain tissue biobank in a small European country:

> Yes, we don't have a problem with it [recontacting/re-consent]. We send a letter, we never call the people. Sometimes the letter comes back saying that person has moved away or that the person has passed away, or they come back and they say no problem. Depending on the answer—sometimes people move from their house to a nursing home or an elderly people's home—and then we try to contact the physician. What happens is that what we put in our original papers is a sentence that clearly says that 'if you move to a different address, please let us know'. I must say that 80% of the people that are mentally capable send us a card, and then we send them a new donor card. Sometimes people change the GP [general practitioner] or they go to a nursing home, and they also send us a card. They say 'I have moved and now I have a new doctor, would you please send me a new card'. So my experience is that those people who sign up as brain donor, they really consider it very carefully and very long and they are motivated to have their autopsy. They make sure that either themselves or the family will contact the bank and give any changes of address or other information (#30 IS, Europe, medicine, bioethics, government, R/A/S).

Arguments against new informed consent to each new study   More than half of the respondents from Europe and from all other regions outside North America and a substantial minority from the US ruled out this type of consent. Most of them adduced practical reasons, while acknowledging that it is ethically acceptable or even superior to other forms of consent.

> [Option (a), i.e. new informed consent] is an appropriate mechanism. The reality is you are not going to get many approvals with this because it is onerous and expensive to do [I rule out (a)] 'from the practicability point of view, the resources it takes. It is an option, but it is not practical (#23 US, medicine, humanities, genetics, university, R/A/S).

Many respondents stated that re-consenting is not feasible, too expensive and too complicated, especially in a developing country.

> [Recontacting would be possible] if the conditions of a developed country were present. Here [in Africa] it would be very, very complicated. Unfortunately, that's how things are, but one has to accept that this is one of the difficulties (#56 IS, Europe/Europe and Africa, medicine, life sciences, government, hospital, S/A/R).

> Societies are too mobile; people are running all over the place, even in organized societies like Britain and Switzerland. Even there is it [recontacting] tough to do (#98 US, philosophy, medicine, genetics, medicine, university, R/A/S).

As it implies many practical problems, re-consent was identified as a burden for researchers and repositories:

> (A) [new informed consent] is ethically fine, but repositories want to avoid recontacting people if they can, it is ethically acceptable but it is probably not desirable from those researchers' standpoint (#41 US, bioethics, law, university hospital, R/A).

> I thought (a) was ideal but too cumbersome. No, I would not rule it out if researchers are willing to do it, I just don't think it should be required (#07 US, social sciences, bioethics, university, O).

The evaluation of burdens and feasibility was based on personal experience or on reports in the literature. Many respondents opposing this type of consent had used or stored samples and described their experiences expressively:

> Recontacting is a headache (#15 US, genetics, university, R/A/S).

Re-consenting for each new study was not only ruled out because of the burden to the researcher, but also because it negatively affects the scientific value of the research:

> Practically it [new informed consent] is not going to be possible. You are not going to be able to recontact everybody and get their explicit consent, you just lose so many. That has been the experience in any kind of recontact (#80 US, genetics, bioethics, philosophy, theology, university, R/A/S).

In addition, the burden resulting from this form of consent is perceived to be important not only for the researchers, but also for the research participants in the US (#96 US, genetics, medicine, social sciences, university hospital, R/A/S), as well as in other countries, "because it would be invasive and intrusive" (#72 IS, Europe/ Africa and Europe [nationality: Europe], medicine, S/A) and it could amount to "a sort of harassment":

> I would rule out (a) if this is a repository that will be kept a long time and there will be lots of studies. Recontacting people 10 times a year would be bothersome and a sort of harassment. It is on the border of an ethical concern and a practical one (#27 US, philosophy, bioethics, government, R/A).

Re-consent is judged to be not in line with the wishes of participants:

> In general I think (a) [re-consent] is not a good approach. Because it requires being able to find people again and to rely on people ... if I was advising investigators and subjects how to design a study I would advise them: ask for as much permission as you think you would like to do to begin with. Because sometimes you might not be able to find people for all sorts of reasons that has nothing to do with their wishes or interest for participation. And as a subject I would say: I am moving around and I would like them to find me so that they can tell me what is going on but what if they can't find me? I would [not like the study] to stop because of that (#59 US, medicine, bioethics, government, R/A/S).

Finally, one participant opposes re-consent because he does not trust researchers to do it correctly:

> My concern with a [new consent] is that I don't know whether I always ... trust researchers to do what they say they are going to do (#70 US, genetics, social science, bioethics, university, R/A/S).

Arguments in favor of presumed consent to future research projects   The following description of presumed consent was given in scenario A: "For each new study unrelated to colon cancer, participants will be informed and offered a chance to opt out."

This type of consent is the option least often preferred and, together with new informed consent, most often ruled out by the respondents from our study. The few interviewees in favor of it referred to its efficiency and to its advantage of not requiring an active response of the sample donor as it is the case with re-consent:

> Giving people the possibility to opt out as in (b) certainly minimizes the effort it will take on the part of the researcher. ...I think that allowing for each new study unrelated to colon cancer [to take place if] participants will be informed and given a chance to opt out is a more efficient way of accomplishing the same thing. Some would argue that the researcher does need to do the active part of it, but I think about cost effectiveness and effectiveness on all sides and allowing people to opt out is probably the better way to do (#70 US, genetics, social science, bioethics, university, R/A/S).

Some interviewees who preferred a different option, but found presumed consent second best, mentioned conditions necessary to make this consent ethically acceptable. One respondent suggested that presumed consent is acceptable only for certain types of research topics:

> (B) [presumed consent] might be acceptable depending on what the new study unrelated to colon cancer actually was. If it is very unrelated (b) might not be acceptable (#44 US, bioethics, social sciences, A).

Another respondent draw attention to the obligation to obtain—at the beginning of the first study—written consent of research participants to the practice of presumed consent that will apply for future research projects:

> I think that it [presumed consent] is ok but again I think the consent process needs to … really encourage them to read the website/newsletter. Ideally we would think that if people get a newsletter they read it but I do think that part of the consent process should say: more studies will be planned we will use your samples for these studies unless you tell us otherwise and the way you find out what other studies are being planned is to check the website periodically and read the newsletter. So I think that should be emphasized in the consent form and then I think (b) is fine (#07 US, social sciences, bioethics, university, O).

Arguments against presumed consent to future research projects

Presumed consent encountered opposition mainly on ethical grounds. Respondents objected that it is "unethical" (#41 US, bioethics, law, university hospital, R/A). It is said not to respect autonomy adequately on grounds that it is an invalid substitute for consent.

> (B) is an opt out option that is not good, informed consent is not a matter of opting out … it is a poor substitute for informed consent (#20 US, law, philosophy, university, R/A).

> I am not sure that (b) [presumed consent] is coherent because it is not a consent. I would not approve of (b) … (b) is dangerous. I don't like presumed consent, don't call it presumed consent, just say we are going to do it. [It is an] improper use of the term consent (#95 US, bioethics, philosophy, theology, university, N).

In contrast to consent options that are judged ethically acceptable but ruled out for practical reasons, presumed consent is criticized on all levels: it is said to be incoherent, "ethically dubious" (#71 US, Europe, law, philosophy, bioethics, university, N), dangerous and impractical. Respondents against presumed consent are convinced that it is not a reliable means to assure that research participants' wishes are respected:

> I have a notion that we cannot expect too much from people. Informing people of what is going to happen? Forget it! Nobody is informed, nobody reads. People don't care about organ transplantation, [… and even less about] genetic information. We are asking too much. That's more a lip service to our ethical conscience and not a real solution to the ignorance of people (#62 IS, Europe and No. Am/Europe, medicine, law, university, R).

Instead, it is dangerous, because the absence of a response might be interpreted falsely as consent.

> (B) [presumed consent] I think is almost unacceptable because you cannot really be sure that people got your newsletter so the opt out option is not good enough in this scenario (#39 US, Europe, philosophy, bioethics, government/university, R/A).

Respondents are opposed to the use of an unreliable standard of consent because research involving biobanks implies risks for sample donors. Hence, more "responsible" ways are needed to make sure that research subjects do not participate in studies against their will.

> Presumed consent is not acceptable. ... I don't think this is a responsible way to conduct research. So if it is going to be used for anything else they should come back each time and tell them what it's for (#09 US philosopher, catholic theology, university, N).

> For research I think explicit affirmation of one's willingness is always better than just opting out. I mean what does no response mean? It is very amorphous. ... I think affirmative consent is far superior to not opting out. And it is also much clearer (#54 US, bioethics, protestant theology, philosophy, university, R/A).

In addition, interviewees oppose presumed consent because they fear negative consequences not only for the repository and researchers, but also for participants. Presumed consent imposes an excessive burden on participants:

> It places too much burden on the participant to keep up with the information and there is too much of a possibility that they are not being adequately notified (#01 US, medicine, bioethics, university, N; reason why presumed consent should be ruled out).

The negative consequences for repositories and researchers are framed in terms of practical difficulties, burdens and excessive costs.

> I think it [presumed consent] would be difficult in practice. These repositories are long term ventures and the cost of keeping track of people would be disproportionate (#82 IS, Europe, natural science, A/R).

> (A) [new informed consent] and (b) (presumed consent) are both viable kinds of alternatives. I can see their attraction. Logistically obviously they are quite difficult and very time consuming, very expensive. ... logistically I think they are nightmares but one could do [them] as long as you have sufficient funds (#36 US, medicine, genetics, government, R/A/S).

While many oppose presumed consent because they find this type of consent insufficient, others consider it "overkill" (#92 US, law, medicine, genetics, university, R/A). These respondents think that one-time-consent at the beginning is sufficient, either in form of general or multilayered consent. Any later contact with research participants to give them the possibility to change their minds is judged ethically unnecessary, although it might still be good practice:

I think that (b) [presumed consent] is a possibility, but it does not seem as necessary, you can send newsletters to everybody about what is going on and those who are unhappy with it could use this as an opportunity to opt out. I think it is good in general to let people know what is going on but that is different from the need for consent (#26 US, medicine, social sciences, government, R/A).

Multilayered consent   In scenario A.4, the following form of multilayered consent options was proposed: "The consent form should be modified to allow the participants to choose from the options mentioned above" (new informed consent to each new research project, presumed consent, general consent).

Multilayered consent was the type of choice that received the second best ranking in the US as well as in Europe: one third of respondents consider it their preferred option. However, in the remaining regions, both new informed consent and general consent were chosen as the best option more often than multilayered consent. Overall, multilayered consent solicited strongly opposite reactions: interviewees were either very much in favor of it or categorically against.

Arguments in favor of multilayered consent   Respondents from different cultural contexts indicated that multilayered consent is the best choice because it is the only way to respect the autonomy and the wishes of different participants in a global, pluralistic society in which "several moralities" exist:

But [some] people [may] not agree, and some people may agree to store the sample and use it for all researches related with colon cancer ... I think of this [in terms of respect for] autonomy. There are different kinds of people; we have several ideas about different topics ... So you can have different ideas about the same storage of sample […I am] considering the several moralities we have (#65 So. Am., life sciences, genetics, university, A).

I think (d) [multilayered consent] is the best. It ... asks participants how much in essence they care about the range of things of not knowing what it might be used for. And different participants will care differently and some don't want to be burdened by having the doorbell ring every time a new study has come up. And some will care very much. They would want to ask 'what is the new study? I am not really sure,' so they can self-select this way (#16 US, philosophy, bioethics, university, N).

Others explained further that informed consent implies respect for autonomy, which in turn means that choice between different options must be offered.

(D) [multilayered consent] is best ... The informed consent must give participants a choice (#34 IS, Europe, medicine, genetics, biobank, S).

I think that (d) [multilayered consent] is the preferable one because it makes more things explicit [in terms of] objects of choice (#02 US, philosophy, catholic theology, university, N).

I like (d). It just gives the person the option to decide whatever they want (#25 US, philosophy, government, N).

One respondent added that multilayered consent is best because it has educational value: in order to make decisions in a truly autonomous way, research participants need to be able to compare different options.

> [I prefer (d), i.e. multilayered consent] because it is a menu consent that really allows individuals to not only choose but because it gives them the different options; it opens their minds to considering and comparing the options. So I think a menu consent is valuable not just because it allows you choice but because actually lining up the different options has almost an educational value. It helps people think through this (#39 US, Europe, philosophy, bioethics, government/university, R/A).

Although respondents in favor of multilayered consent admitted that it is more complicated than general consent, they are convinced that "it is manageable" and "not too confusing" (#73 IS, No. Am./Asia, life science, government, S) for sample donors and researchers in different parts of the world. Others pointed out that the possible burden for researchers can be reduced by using technical means such as electronic coding (#12 US, Africa, genetics, university, R/A/S).

For several interviewees the ethical reasons in favor of multilayered consent such as the importance of respecting different wishes of participants outweighs the possible negative consequences of this type of consent for the researcher and even for the scientific value of the study:

> What are the realities: would it bias the study or not is irrelevant. [What's important is] what the participants said you can use their resources for (#12 US, Africa, genetics, university, R/A/S).

The burden is judged ethically irrelevant by several respondents from the US, perceived as a country that has enough resources for research.

> (D) [multilayered consent] is the best, I mean it would be a nightmare to administer but why not (#52 US, social science, bioethics, university, R/A).

One non-scientist admitted that he is not able to understand possible disadvantages for research and therefore does not see how these problems could outweigh the advantages in favor of multilayered consent:

> Again being a non scientist I am not understanding how this [multilayered consent=d] can really limit valuable information that can be gained. I do think that (d) is better than (c) [general consent] because they are just giving an open ended consent as I understand it in (c) and in (d) they at least have a chance to address it (#09 US philosopher, catholic theology, university, N).

Interestingly, some believe that what is often called a burden is in fact helpful for the research process. Hence, these respondents chose multilayered consent because of its benefit for the research.

> The concern I am burdening the researchers is somewhat misguided … I think that of course talking with patients … burdens research. On the other hand, having people have a say in the research process is in fact something that strongly supports research. So …

honoring and acknowledging and giving them control and a say over what is done with samples they donate ... is actually a good thing and is good for research, so [I am] far from viewing it just as a burden. It is a question about giving them an appropriate array of choices about what kind of research they are involved in or not (#92 US, law, medicine, genetics, university, R/A).

Another beneficial effect for the research is the perception that multilayered consent may allow to enroll more research subjects:

(D) [multilayered consent] allows for more data to be used in the beginning if some participants are hesitant so if the researchers have enough staff capability for that there is nothing wrong with doing (d) (#64 US, medicine, bioethics, university hospital, N).

Several respondents pointed out that multilayered consent is the best option only under particular conditions. First, adequate multilayered consent should provide more information as it does in its present form. The consent form should explain in more detail the future uses of the samples and contain more choices in terms of disease categories.

(C) [general consent] is acceptable and (d) [multilayered consent] preferable, [but] it would be desirable to specify what those other purposes are or might be (#71 US, Europe, law, philosophy, bioethics, university, N).

I think (d) [multilayered consent] is ideal as long as there are more categories (#07 US, social sciences, bioethics, university, O).

One respondent sees multilayered consent as morally unproblematic only if sufficient efforts are undertaken to make sure that research participants have understood what the choices mean:

I think (d) [multilayered consent] is fine provided people could figure it out, I am just not sure it is the most practical way to proceed. [...It is] ethically probably unproblematic presuming that you work hard making it clear what these choices mean. So from a moral point of view if you put a lot of effort in the consent process (d) could work just fine, logistically it is problematic (#44 US, bioethics, social sciences, A).

Another respondent wonders whether recontacting, included as one of the multilayered options, is practical. He proposed to limit the list of options and to include only those options that are known to be feasible (#28 US, bioethics, medicine, government, N).

Arguments against multilayered consent Multilayered consent was ruled out by almost half of respondents from Europe, one third of respondents from the US sample and less than one fifth of respondents from elsewhere. It was ruled out for similar practical reasons as new informed consent to each new study. Respondents considered that the options in the multilayered consent are too "complex" (#99 US, medicine, genetics, university hospital, R/S) and "impractical beyond description" (#98 US, philosophy, medicine, genetics, university, R/A/S). Without knowing each other's answers, a remarkable number of interviewees used the term "nightmare" to describe this type of consent.

I would dismiss (d) [multilayered consent]; (d) is respectful of autonomy and decision making authority and all that but it is really unclear how you would ever manage this. It creates kind of a managerial nightmare since you have patients/different donors choosing different ways. And then, how are you going about recontacting and kind of managing that across people especially if you have families involved (#10 US, humanities, law, university, R/A).

It should be noted that several respondents (e.g. #66 IS, Europe, law, consultancy, A/R and #82 IS, Europe, natural science, A/R) are against multilayered consent in the form proposed in scenario A.4 not because it is impractical as such, but because it offers one "disproportionately expensive" option: the option to ask for re-consent to each new study.

I would not favor (d) [multilayered consent] because that would be administratively extremely expensive and difficult to manage, if you have to go back to your sample [donors] and say 'fine, we collected for colon cancer but now we want to look at Alzheimer disease'. It would just be impractical and it would make further research disproportionately expensive (#82 IS, Europe, natural science, A/R).

When discussing impracticability, respondents mentioned undesirable consequences for researchers and the efficiency of the research projects.

One negative effect described by several respondents is the lower scientific value of the biobank and different research studies resulting from heterogeneous consent for the use of the samples and information.

[W]e need to develop research and if you do public health population-based research, you cannot just pick up a couple of samples 'yes' and a couple of samples 'no'. The random sample is random if you have the whole [set of] samples (#57 IS, No. Am./Europe, medicine, university, S).

Multilayered consent doesn't work very well because then you have different rules for different people. It is possible to do it but I think you then end up with samples that are degrading at different rates. Let's say that people with Asian backgrounds degrade a lot faster than other people, and women faster than men or vice versa. It is not a very good research design to ask at enrolment 'which one do you choose?' (#68 IS, No. Am./Europe [nationality: No.Am.], life sciences, consultancy R/A).

Furthermore, the complicated management of multilayered consent leads to "poor science":

I don't think [options] (a) [new informed consent], (b) [presumed consent] and (d) [multilayered consent] are broad enough, and they cause a lot of back and forth relationship with the patient. I think it is poor science in that regard (#63 US, bioethics, medicine, catholic theology, university, A).

Respondents against this type of consent also argued that it is "without any added ethical value" while being "overly burdensome" (#33 US, bioethics, medicine, government, A).

> I think (d) [multilayered consent] is very cumbersome and actually places a burden on both investigators and subjects that is unnecessary (#95 US, bioethics, philosophy, theology, university, N).

Respondents against multilayered consent judge it "unnecessary" because they have a different understanding of the scope of respect for autonomy. The latter does not imply an unlimited right to choose. It is sufficient to grant the right to accept or to refuse participation in future research projects involving a biobank. In this context, one respondent gives a caricatured example to show that the right to choose for research participants is always limited:

> We didn't think also that it was justifiable for people to choose what researcher [could use them]—they were happy with people in [town 1] and [town 2] using it but not with people from [town 3] (#68 IS, No. Am./Europe [nationality: No.Am.], life sciences, consultancy R/A).

However, not all interviewees against multilayered consent consider the right to choose limited. In line with respondents favoring multilayered consent, several respondents opposed to it acknowledged the ethical value of offering choices, namely the respect for distinctive preferences of different participants. However, in contrast to those in favor of multilayered consent, opponents do not think that these ethical advantages outweigh the disadvantages for research.

> Ideally is option (d) but it may make research impossible (#53 IS, Europe, medicine, public health, Int. Org., R/A/S).

Furthermore, one respondent pointed out that it is ethically inappropriate to make promises that might not be kept, a possible result of the logistical complexity of managing the different degrees of consent.

> I think it might even be better just to do (c) [general consent] right upfront, (d) [multilayered consent] you might be promising something you can't deliver, even though in the ideal world that might be nice ... You could do a menu consent if the database wants to do that, but then you are really running several databases rather than one ... It adds a great deal of complexity (#80 US, genetics, bioethics, philosophy, theology, university, R/A/S).

Respondents favoring and opposing multilayered consent disagree about the empirical question whether offering more options to research subjects is beneficial or not. While those defending multilayered consent think it has educational value and increases research participants' capacity to make a decision in line with their values, opponents are convinced that multilayered consent decreases this capacity because more choice adds "confusion" (#47 US, law, social science, government, A).

> If you give too many options they [participants] get confused ... Patients should be able to say 'only for colon cancer' or '[also for] other diseases' (#78 IS, Asia, life sciences, government, S/A).

This view, that two choices are preferable because more choices increase confusion, was found across different regions in the world, including the US:

(D) [multilayered consent] is what I find most troubling because it requires a huge amount of energy to evaluate those choices and to consider a decision ... I think it puts an extraordinary burden on subjects ... And that set of choices [is useless] unless you were imagining having a thirty-five-minutes-discussion how you should make this choice. People would not know how to make it and what they choose might not even reflect what they actually believe. And if you are interested in trying to help people at some level accomplish what they believe in I think that three makes it more difficult ... I think the multiple choices create a level of complexity and confusion .We would need further research so that it is clear that people understood everything (#59 US, medicine, bioethics, government, R/A/S).

Hence, multilayered consent is seen by these respondents as a way to protect researchers without offering real choice to participants.

I think (d) is again a way to try to protect the researcher rather than to permit a real choice for the participant (#15 US, genetics, university, R/A/S).

To buttress their argument, several respondents referred to empirical studies on the benefit of providing choices. According to the interviewees, these studies show that it is not in the interest of individuals to be confronted with too many options because they add too much complexity and make it more difficult for people to identify the choice that best agrees with their preferences.

I also think there is just something about choices ... You know from choice analysis and decision making that if you give two options for instance, then throwing in a third option will degrade their ability to make a good choice. So giving people these kind of options is not good (#10 US, humanities, law, university, R/A).

Interviewees from different continents mentioned the fact that patients lack the education to understand multilayered consent:

I also think that (d) would be the fairest of them all. But not all people in [my country in So. Am.] have a university degree. If you give options to choose on complicated matters, you may not actually get the most correct answer. And I would ask them explicitly to agree. [This] is better than asking them a choice, of opting for one of the alternatives that may be not clear to them (#60 IS, So. Am, pharmacology, medicine, government, S/R).

[I]t doesn't make much sense to put a list of diseases, people don't understand what the words mean without even talking about the ramifications (#62 IS, Europe and No. Am/ Europe, medicine, law, university, R).

As these quotations and the following show, several respondents referred to their personal experience to explain their view that research participants lack sufficient education to understand information related to different choices and that more explanations will lead to even more confusion:

I want to make a point here, and I am talking based on my experience. Whenever we [discuss] with participants about studies, we find that they are not well educated, they are not well informed of the details of the fate of their samples. They don't know what will happen really for these samples later on ... Some will give you the sample without knowing what will happen to it. The more you involved the participants at many levels of the work; the more they will get distorted. For me, I always try to find alternatives that will minimize confusing the participants' minds (#93 IS, Europe/Middle East, biology, genetics, biobank, S).

2.3 Notable similarities and differences between subgroups of respondents

Firstly, it is worth mentioning that defenders of classical informed consent for each research project involving biobanks are found in all regions of the world: about half of respondents from both the US and the international samples do not think that general consent is preferable. They consider general consent or presumed consent to contradict the definition and essential meaning of informed consent and therefore deny that it is an ethically acceptable substitute for it. New informed consent and multilayered consent were considered closer to informed consent, with regional differences existing for both. New informed consent is preferred by merely a few respondents from the US and Europe, whereas a substantial minority of interviewees from other regions considered it the best form of consent in scenario A. Several explanations for these differences emerge from the interviews.

The most important seems to be a greater fear of abuse outside Europe and North America, in particular among members of indigenous groups. In these regions, new informed consent is viewed as the most adequate means to prevent abuses.

Interestingly, although not the preferred option of most respondents from the US, new informed consent is less frequently ruled out in this country (by about one fourth of respondents) than in Europe and the remaining regions, where more than half of the interviewees oppose it. This indicates the extent to which the classical form of new informed consent is respected in the US. It remains an unpractical, but still acceptable option in this country, whereas other regions seem to be more prepared to reject it outright.

In contrast to new informed consent, multilayered consent was an option that ranked low among respondents from outside Europe and the US. It seems to be a typically US American conviction, but also defended in Western European countries, that a large number of choices is the solution to consent problems in the context of biobanks. Perhaps these findings are explained by the influence of North American guidelines, several of which recommend multilayered consent, on some US respondents. In light of these recommendations, it is remarkable that more than half of US respondents favor general consent, especially interviewees from governmental research institutions and those who have worked with samples.

Our results show also that general consent is the preferred option of almost all respondents from Asia and about half of the interviewees from Europe and the US, whereas it cuts no ice with most respondents from Oceania, Africa, and South America. It may be that in many developing countries, trust in institutional safeguards is lower, and respondents doubt that privacy will be protected and that ethics committees will allow only "ethical" studies to go through.

Finally, it turns out that respondents who have used or stored samples differ from respondents without such prior involvement. The former preferred general consent more often than the latter. They chose new informed consent less often than non sample users and indicated practical reasons, especially the disproportionate burden of these types of consent. For similar practical reasons, a substantial minority of them criticized multilayered consent. These findings can be explained in several ways: on the one hand respondents who used samples are in a better position to know about burdens they may have experienced themselves. On the other hand, as admitted by several respondents, researchers' attitudes reflect their own interests to avoid inconvenience for research, and they may see the balance with patients rights somewhat differently.

The second particularity of interviewees who used samples seems to be a remarkably low estimate of research participants' capacities to understand research issues and consent choices. Again apparently based on personal experience, several respondents who used samples oppose multilayered consent, as well as new consent to all future studies, because it is said to overburden and confuse research subjects. Again, it cannot be excluded that this estimate is influenced by personal interests to limit the inconvenience of convoluted explanations to research participants.

3. Discussion and conclusions

3.1. Reasons for agreement and disagreement

Construed consent   Although acceptation for different forms of consent varies widely, at one point the ethical sensibilities of most respondents are similarly hurt: consent should not be construed. The reason why, with few exceptions, the policy in scenario A.3 was rejected seems to be that in this case most think that consent is misconstrued because participants are misled. It is acceptable for some respondents to extend the use of samples if consent was vague and use for different diseases was not excluded. But if research participants consented to the use of their samples for research on colon cancer, most find their wishes must be respected as such. It should be noted that this scenario has distinct features and should not be confounded with the situation presented in scenario D.2 (see next chapter), where the question was framed differently: should it be allowed to use hospital sample collections for research if this purpose had not been mentioned in the original consent form? In scenario A the collection of samples has started recently, the contact with participants is ongoing through their physicians because changes in lifestyle and health status will be reported. Asking patients and family members whether they would also consent to research on other diseases is considered essential by most interviewees if a research biobank is about to be established.

What is informed consent?   Respondents hold different views about what constitutes ethically acceptable informed consent. Defendants of the classical form of consent for each research project are found in various parts of the world. This apodictic deontological attitude towards consent is incompatible with the view of others who accept general consent to future uses because "future uses for medical research" is considered sufficient information allowing research participants to make an informed decision. In traditional research ethics, the fact that research participants find general consent acceptable does not automatically make it so. The traditional approach in research ethics is somewhat paternalistic. If research is too dangerous, patients' or subjects' consent is not sufficient to allow participation. Similarly, if subjects agree to insufficient consent standards, one might be obligated to hinder them from forfeiting their rights. To bridge the gap between the existing differences successfully, an international consensus would be needed as well as statements with sufficient authority to compete with the more than fifty year-old tradition started with the Declaration of Helsinki.

Balancing of research efficiency with autonomy rights of research participants Only a few respondents were almost oblivious to consequences such as increased costs, burdens and diminished scientific quality of research. For them, autonomy rights always trump such negative consequences. For the majority of interviewees however, consequences do matter and autonomy rights are judged according to less absolute standards. For these respondents, general consent respects research participants' autonomy sufficiently. Therefore, even if higher standards of consent such as new informed consent or multilayered consent respect autonomy to an even greater extent, costs, risks and burdens will shift preferences towards general consent in its broader and narrower varieties. Since most interviewees referred to positive or negative consequences in favor or against various types of consent, it is obvious that disagreement can result from diverging assumptions about these consequences.

Distinct assumptions about consequences Firstly, disagreement is related to regional variations in anticipated consequences. Although the assumption "the bigger the country, the harder it is to do it" (#98 US) might be too simplistic, recontacting can be unfeasible in the US where "50% of the population moves every year" (#33 US), as well as in developing countries because the costs are prohibitive, whereas it could be a realistic option in a small European country where many patients stay with the same treating physician for their entire life. Regional variations are also evident as regards the perceived risk of abuse. Societies and countries show much variation as regards legal protections, the functioning of ethics committees, and factors influencing the risk of genetic discrimination, such as for example access to health insurance.

Secondly, disagreement is the result of diverging assumptions about consequences that are currently uncertain or difficult to ascertain. Respondents hold different views about the wishes of research participants: some think that most research participants prefer recontacting, whereas others think that most research participants would rather not want to be bothered again later. It is uncertain whether and how the type of consent would influence the participation rate in a given study. Some respondents assume that with general consent more research participants will refuse participation than in the case of multilayered consent. Does multilayered consent cause confusion of research subjects or is it beneficial for the choice process? Some empirical studies exist and were in fact cited by several interviewees, in particular studies showing the large acceptance of general consent in several countries (Wendler 2006). More empirical data on these issues might help to solve some disputes on consent, although these questions are likely subject to regional differences as well. In some developing countries, fear of abuse might cause a high rate of refusal to participate in studies, which is not the case in Europe (Hoeyer 2004).

Disagreement is further due to cultural differences related to the importance of choice as a basic component of respect for autonomy. However, this predominantly US American view of autonomy may be an oversimplification and anyway subject to change. Our results show this view to be shared by only a minority of US respondents. One of them might be right when he diagnosed that attitudes are changing in the US:

> But I think there are illusions about cost and practicality that start to shape some of the dialogue. I have sympathy for the geneticists (#98 US, philosophy, medicine, genetics, university, R/A/S).

3.2. Contributions to the current debate

The answers of the interviewees reflect the various positions adopted in guidelines. However, it is notable that one consent type proposed by the Human Genome Organisation (HUGO 1998), namely presumed consent, is clearly favored less often than the others. It is said to be the least compatible with the meaning of informed consent because no control exists on whether research participants have received the information or not. On the other hand, general consent, which is rarely defended by guidelines, is an acceptable option for half of the interviewees. This is in line with numerous studies that have shown the wide acceptance of general consent (Wendler 2006). Why do a substantial minority of interviewees, especially from outside Europe and the US favor new informed consent for future studies on samples stored in a research biobank? One explanation might be that respondents in our study, ethicists and scientists, felt obligated to some degree to provide "ethically correct" answers. However, this hypothesis is not convincing because it does not explain regional differences. Why would respondents from regions outside Europe and North America be more inclined to "ethical correctness" than those from these continents?

The aim of international guidelines is to create uniform regulations for two main reasons: to assure adequate standards of protection of research participants worldwide and, second, to permit international collaboration. As our results show, regional differences of attitudes exist. They are at least partly justified by regional differences of risks and benefits and therefore difficult to eliminate. Any international policy should take into consideration the different contexts existing worldwide and propose a regulation that protects the most vulnerable. Hence, one option to integrate different regional attitudes could be to choose the highest standard of consent in order to protect the most vulnerable populations in countries with a high risk of abuse. However, critics might say that this leads to high costs and burdens which could be avoided in many countries where general consent is acceptable to the population and to ethicists because protections are considered sufficient and, consequently, the risk of adverse effects resulting from biobank studies might be relatively low.

Another way to prepare the ground for uniform regulations could be to work on practical methods to ensure data and sample security worldwide. Furthermore, efforts to create universal standards for trustworthy research ethics committees and to provide poor countries with the means to develop these structures would facilitate protection. If there is enough evidence that trustworthy ethics committees and feasible data security techniques exist to protect research participants even in developing countries and in indigenous populations, this could increase trust and make a moderate version of general consent acceptable. However, the issue of international guidelines about consent to biobank research remains challenging. Our results show that improved data security will not change the attitudes of two groups of individuals: those who are generally risk-averse and not open to arguments based on consequences and those who hold absolute views on the nature of informed consent and the primacy of research participants' autonomy.

Bibliography

American Society of Human Genetics (1996), "ASHG report. Statement on Informed Consent for Genetic Research," *Am J Hum Genet* 59, 471-74.

Annas, G.J., Glantz, L.H. and Roche, P.A. (1995a), "The Genetic Privacy act and Commentary, unpublished model law, February," cited in Weir, R.F. (1999), "The ongoing debate about stored tissue samples, research, and informed consent," Commissioned Paper in National Bioethics Advisory Commission (1999), *Research involving Human Biological Materials: Ethical Issues and Policy Guidance, Report and Recommendations, vol. 2*. Rockville; F1-F21.

Annas, G.J., Glantz, L.H. and Roche, P.A. (1995b), "Drafting the Genetic Privacy Act: science, policy, and practical considerations," *J Law Med Ethics* 23: 360-6 [1995b].

CCNE (Comité consultatif national d'éthique) (1995), Opinion and Recommendations on "Genetics and medicine: from prediction to prevention." Reports. No 46-30 October 1995 <http://www.ccne-ethique.fr/english/start.htm> accessed 9 July 2007.

COE (2006), *Recommendation Rec(2006)4 of the Committee of Ministers to member states on research on biological materials of human origin* (Strasbourg, France: Council of Europe).

Deschênes, M., Cardinal, G., Knoppers, B.M. and Glass, K.C. (2001), "Human genetic research, DNA banking and consent: a question of 'form'?," *Clin Genet* 59, 221-39.

Elger, B. and Caplan, A. (2006), "Consent and anonymization in research involving biobanks," *EMBO Reports* 7, 661-6.

Elger, B. and Mauron, A. (2003), "A presumed-consent model for regulating informed consent of genetic research involving DNA banking," in *Populations and Genetics: Legal Socio-Ethical Perspectives*, Knoppers B.M. (ed.) (Leiden/Boston: Martinus Nijhoff Publishers), pp. 269-95.

European Society of Human Genetics (2003), "Data storage and DNA banking for biomedical research: technical, social and ethical issues. Recommendations of the European Society of Human Genetics," *European Journal of Human Genetics* 11, Suppl. 2, S8-10.

Hoeyer, K., Olofsson, B.O., Mjorndal, T. and Lynoe, N. (2004), "Informed consent and biobanks: a population-based study of attitudes towards tissue donation for genetic research," *Scand J Public Health* 32, 224-29.

HUGO [Human Genome Organisation] Ethics Committee (1998), *Statement on DNA Sampling: Control and Access*: http://www.hugo-international.org/Statement_ on_DNA_Sampling.htm>, accessed July 2007.

HUGO [Human Genome Organisation] Ethics Committee (2003), "Statement on Human Genomic Databases. December 2002," *Eubios Journal of Asian and International Bioethics* 13, 99.

Kegley, J.A. (2004), "Challenges to informed consent," *EMBO Rep* 5, 832-36.

Medical Research Council of Canada, Natural Sciences and Engineering Research Council of Canada and Social Sciences and Humanities Research Council of Canada, (1998), Tri-Council Policy Statement: Ethical Conduct for Research Involving Humans (Ottawa: Public Works and Government Services Canada).

NBAC (1999), *Research involving Human Biological Materials: Ethical Issues and Policy Guidance, Report and Recommendations, vol. 1.* (Rockville: National Bioethics Advisory Commission).

Network of Applied Genetic Medicine (RMGA) (Québec, Canada). (2000), Statement of principles: Human genome research (version 2000). Network of Applied Genetic Medicine <http://www.rmga.qc.ca>, accessed August 2007.

UNESCO International Bioethics Committee (2003), *International Declaration on Human Genetic Data* (Paris, France: UNESCO).

Wendler, D. (2006), "One-time general consent for research on biological samples," *BMJ* 332, 544-7.

WHO Human Genetics programme (1998), *Proposed International Guidelines on Ethical Issues in Medical Genetics and Genetic Services (Report of a meeting on ethical issues in medical genetics Geneva 15-16 December 1997)* (Geneva: World Health Organization).

World Health Organization (European Partnership on Patients' Rights and Citizens' Empowerment) (2003), *Genetic Databases, Assessing the Benefits and the Impact on Human Rights and Patient Rights*, Geneva: World Health Organization.

Chapter 6

Consent to Research Involving Human Biological Samples Obtained During Medical Care

Bernice Elger

1. Introduction

In this chapter we present the attitudes of the respondents from our study towards informed consent to research on samples which have been collected in an entirely clinical context for the purpose of diagnostic testing. It is well known that in most countries hospitals and institutes of pathology store a huge amount of human biological materials. These samples are important for the patients themselves and help secure, reconfirm or later adjust or modify diagnosis. In addition, they are of immense value for research. However, in many cases consent to the use of these samples was vague or limited to clinical purposes only. Should researchers be allowed to carry out medical research using these samples and if yes under which conditions?

1.1 Informed consent to samples obtained during medical care and consent to existing collections: the challenge

The ethical and legal questions related to the use of samples obtained in an entirely clinical context without consent specific to research are not completely different from those arising from secondary research use of samples taken for storage in a biobank established primarily for research. However, this situation displays particular features in three respects. Firstly, if samples are taken in the context of a research biobank, donors have been informed at least once that samples will be used for research. This is not the case if samples have been taken for diagnostic purposes only. Secondly, hospital patients see a physician in circumstances in which they are particularly vulnerable. They suffer from a health problem serious enough to warrant hospitalization. Is it appropriate to approach them for consent to research involving their samples taken for diagnostic procedures, and if yes, when and how should this consent be obtained to avoid unduly burdening patients and creating administrative nightmares for hospitals? Thirdly, even if one decides on an adequate policy to obtain routine consent to research on samples taken every day in hospitals, one is still left with the problem of already existing collections. These can be very large and have often been taken many years ago in a different ethical and legal climate where questions about consent and secondary uses have not been addressed. Although existing collections have been used for research without consent during many years in the past, public awareness has changed. Repeated outcries in the press reflected the outrage many people felt when they discovered that hospitals and researchers stored and used biological materials from patients, deceased or not, without them ever being informed about such practice, let alone asked for consent.

1.2 Informed consent to samples obtained during medical care and consent to existing collections: published recommendations

Several guidelines differentiate between new and existing collections (COE 2006, ESHG 2003). Standards for consent for research involving already existing collections are lower than those for new collections. This leads to the risk that new collections are established without appropriate consent and later used extensively under the exceptional regulations applying to existing collections. However, only some guidelines stipulate in detail that all biological material collected after the implementation of these particular guidelines will not be subject to exemptions (SAMS 2006). Others limit themselves to a general statement indicating that "[w]henever possible, information should be given and consent or authorization requested before biological materials are removed" (COE 2006 Art. 22 paragraph 2).

Guidelines agree on one point: collections established for other purposes may be used for research without consent if the samples and information are irreversibly anonymized[1] (ESHG 2003, SAMS 2006, COE 2006, UNESCO 2003). Although it is never stated explicitly, this seems to imply that samples and data may be anonymized without the permission of the donor. Few guidelines draw attention to possible adverse consequences of irreversible anonymization. The European Society of Human Genetics (ESHG 2003 p. S9) stresses that the "decision to strip samples of identifiers irreversibly needs careful consideration. The benefit of having unlinked anonymized samples is to secure absolute confidentiality and thereby allows further use of the samples. However, retaining identifiers, while requiring further consent from the subject, will permit more effective biomedical research and the possibility of recontacting the subject when a therapeutic option becomes available."

Most recommendations have provisions indicating under which conditions reversibly anonymized material may be used without consent. These conditions vary. The least strict regulations are proposed by geneticists for a category called "old collections." The criteria that make a collection old are not specified. The only condition for use is ethics committee approval.

---

1 The only exception are the WHO guidelines from 2003 which adopt a standard of "proportional" or "reasonable" anonymity: "Research using archival material, such as pre-existing health records, specific health disorder databases or physical samples that have been retained—for which no specific consent has been obtained—is only permissible if the material and information derived from it is anonymized, and there is no prospect that research results will be used to identify the sample sources at any future time" (WHO 2003, recommendation 10).

Old collections should be regarded as abandoned and therefore useable for new research purposes as long as ethics committee approval is obtained (ESHG 2003, p. S9).

For existing collections that are not "old," the ESHG states that in addition to ethics committee approval, the following conditions apply: (a) recontacting subjects to obtain new consent for new studies is "impracticable"; (b) an appropriate ethics review board has given its consent for further use of the samples; (c) the ethics committee bases approval on the notion of minimum risk for the donor.

The NBAC report (NBAC 1999) includes a similar strategy permitting the use of existing collections, called "waivers." New consent is generally required for each future study, but exceptions are allowed if several conditions are fulfilled, in line with federal regulations in the US (45 CFR 46.116(d), cited in NBAC 1999, p. 66): (a) the research could not practicably be carried out without the waiver or alteration, (b) an ethics committee has approved the "waiver" (c) the research involves no more than minimal risk to the subjects, (d) the waiver or alteration of consent will not adversely affect the rights and welfare of the subjects, and (whenever appropriate), (e) the subjects will be provided with additional pertinent information after participation. NBAC adds that criterion (e) "usually does not apply to research using human biological materials." The reason given by NBAC is that this criterion "might be harmful if it forced investigators to recontact individuals who might not have been aware that their materials were being used in research" (NBAC 1999, p. 70). Conditions (a) to (c) are similar to those stipulated by the ESHG whereas (d) specifies more detailed requirements. Interestingly, for research involving biological material, the NBAC adds another condition that could contradict the restrictions posed on the criterion (e) concerning the possible harm of recontacting. Even if a waiver is granted, "it is still appropriate to seek consent however, in order to show respect for the subject, unless it is impracticable to locate him or her in order to obtain it" (NBAC 1999, note 8, p. 66). The NBAC explains that if a waiver is granted "subjects would, if possible, be contacted and given the choice of opting out; if they did not respond or could not be found, the sample still could be used because the consent requirement already would have been waived" (NBAC 1999, p.70). On the contrary, if the consent requirement has not been waived, only the samples of subjects who responded and confirmed their consent could be used in the research protocol, but not the samples of subjects who did not respond.

Both the ESHG and the NBAC do not specify the criteria for impracticability to recontact subjects or to carry out the research. The guidelines from the Council of Europe (COE 2006, Art. 22) introduce the concept of "reasonable efforts":

> If contacting the person concerned is not possible with reasonable efforts, these biological materials should only be used in the research project subject to independent evaluation of the fulfillment of the following conditions: a. the research addresses an important scientific interest; b. the aims of the research could not reasonably be achieved using biological materials for which consent can be obtained; and c. there is no evidence that the person concerned has expressly opposed such research use (COE 2006, Art. 22 paragraph 1.ii.).

In comparison to the ESHG and the NBAC, other additional conditions are used: the "important scientific interest" and the lack of "evidence that the person concerned has expressly opposed to such research." While the Council of Europe refers to "scientific interest" only when explaining the condition related to the benefit of research, UNESCO uses an even broader concept. The condition is described as the collection's "significance for medical and scientific research purposes e.g. epidemiological studies, or public health purposes." This exceptional use of existing collections under these circumstances is only allowed if it is compatible with domestic law. Otherwise UNESCO limits conditions to ethics committee approval, albeit asking for the involvement of ethics committees "where appropriate ... at national level."

Some guidelines specifically discuss consent to research on samples obtained during medical care. The HUGO statement (HUGO 1998) proposes a form of presumed consent for the research use of samples obtained during medical care devoid of research aspects. HUGO differentiates between "routine samples, obtained during medical care and stored" and "research samples obtained with consent and stored." According to HUGO, routine and research samples may both be used for research if there is general notification of such a policy, the patient/participant has not objected, and the sample to be used has been coded or anonymized. Research samples obtained before notification about such an opt out policy, may be used for other research if the sample has been *coded or anonymized* prior to use, whereas routine samples obtained during medical care and stored before such notification of such a policy may be used for research only if the sample has been *(irreversibly) anonymized* prior to use.

NBAC stipulates that informed consent to the research use of human biological materials "should be obtained separately from informed consent to clinical procedures" (NBAC 1999, p. 64 recommendation 6) and that it should be made clear to potential subjects "that their refusal to consent to the research use of biological materials will in no way affect the quality of their clinical care" (NBAC 1999, p. 64 recommendation 7). NBAC mentions explicitly that "general releases for research given in conjunction with a clinical or surgical procedure must not be presumed to cover all types of research over an indefinite period of time." In this case, new informed consent must be obtained, unless an IRB waives the consent requirement or "the identifiers are stripped so that samples are unlinked" (NBAC 1999, p. 64 recommendation 8).

In addition, WHO guidelines (WHO 2003) could be invoked to defend a somewhat similar policy based on presumed consent to be implemented in hospitals with regard to samples taken for diagnostic procedures:

> Those who would seek to depart from the practice of requiring active informed consent prior to participation in the creation of a genetic database must justify this position in strong ethical terms. As a minimum the following criteria must be satisfied: (a) a clear, realizable and significant public health benefit must be identified, (b) the widest possible educational program must be instituted among the population that will participate, including an opportunity for public debate (c) strong privacy protection measures must be implemented, (d) individuals must at all times be given the opportunity to refuse to participate, and (e) every stage of the process must be subject to the most stringent ethical scrutiny (WHO 2003, Recommendation 14).

1.3. International recommendations about consent and anonymisation

Many international guidelines do not require informed consent prior to research use of irreversibly anonymized samples or data (e.g. UNESCO 2003, Art. 16(b)). Some nuances are interesting. The Council of Europe specifies that previously expressed restrictions by patients on the use of their samples must be respected:

> Unlinked anonymized biological materials may be used in research provided that such use does not violate any restrictions placed by the person concerned prior to the anonymization of the materials (COE 2006, Art 23).

An international organization of geneticists goes further and considers anonymous genetic data and DNA samples as "abandoned data." Their use and protection is not any more governed by the person who provided samples or data, but by "the processor and/or Principal investigator" who "should be considered as the custodian of these data and should take any steps to protect the data, its storage, use and access" (ESHG 2003, S10).

1.4 Which type of consent is adequate for research involving samples collected in a clinical context (scenario D)?

The issue of subsequent research using samples obtained for diagnostic purposes was addressed in scenario D. This scenario describes a common ethical problem. At present, during hospitalization, patients consent to the storage of their biological materials for future diagnostic testing. Frequently, these samples are interesting for research, but routine procedures for consent to research are lacking. The scenario emulates real life referring to a law on consent. As a matter of fact, several countries have enacted laws or at least issued recommendations such as those of the Swiss Academy of Medical Sciences (SAMS 2006) indicating that informed consent is required for the storage and use of biological samples for research purposes.

In scenario D, biological materials have been taken for future diagnostic testing in a university hospital in an entirely clinical context. The hospital considers implementing routine consent procedures to permit research use of the samples. The vignette situates the hospital in a country in which a law has recently been enacted requiring informed consent for the storage of biological material for research purposes. We were interested to know which type of consent respondents consider adequate for prospective research involving hospital samples. Interviewees were asked to argue in favor or against three proposed alternatives to be used as routine procedure in the hospital. They indicated which alternatives they consider appropriate to comply with the legal requirement of informed consent and whether any of these alternatives should be ruled out. The three alternatives were (1) general consent ("I agree to my stored specimen being used for medical research"), (2) presumed consent ("I agree to my stored specimen being used for medical research provided I have received advance notice and I have not opted out of the study"), and (3) new consent to each use ("I agree to my stored specimen being used for medical research provided I have consented to each use").

In a second part, we inquired about respondents' attitudes concerning consent to research involving already existing sample collections stored in the hospital setting.

2. Results

2.1. Using samples obtained during medical care for research, which is the preferable routine procedure: arguments in favor and against general consent, presumed consent and new informed consent for each research project

More than two thirds of respondents from Europe, about half of respondents from the US, but only about one third of interviewees from other regions indicated that general consent to "medical research" is adequate in scenario D.1. In all three regional groups, about one tenth favored presumed consent. One seventh of respondents from Europe, one fourth of those from the US, but almost half of respondents from other regions said that new informed consent is the most adequate consent type. Indeed, in scenario D this consent policy was generally preferred more often than in scenario A where less than one third of the interviewees from outside the US and outside Europe and less than 10% of respondents from the US and Europe chose this policy as best consent type. Only about half of all respondents chose the same type of consent in both scenario A and D. The differences are explained to a large extent by the fact that respondents from Europe and the US who chose multilayered consent in scenario A indicated either new informed consent or general consent in scenario D where the option "multilayered consent" was not offered. Respondents from other regions changed more often from general consent in scenario A to new informed consent in scenario D than respondents from Europe and the US.

Presumed consent has been approved by respondents somewhat more often in scenario D than in scenario A. If one does not count the handful of respondents who do not agree with any of the proposed consent options in scenario D and propose multilayered consent or mixed forms, one is left with two groups of respondents who hold mutually exclusive views. One group supports general consent and rules out new informed consent and the other group believes that only new informed consent is appropriate and general consent should be ruled out.

Arguments against and in favor of the consent options were essentially the same as in scenario A (see chapter 5) and will not be repeated here. We will report predominantly those arguments that show new aspects or are specific to the clinical context in scenario D.

Arguments in favor of any type of explicit consent   The fundamental concern of respondents independently from the type of consent they defended in scenario D.1 was that respect for autonomy implies that hospital patients are adequately informed about the purpose of storage.

> If consent was given for the use of specimens for a certain specified purpose, it should be limited to that purpose [unless] you ask them right at the beginning explicitly (#21 IS, Europe/Asia, philosophy, university, A).

Interviewees argued that in order to comply with the meaning of informed consent "the consent should be specific, and say "the specimen will be used in research." (#35 IS, Europe/So.Am., medicine, genetics, university, S).

Respondents are concerned about the possibility that hospital patients are put under pressure to participate in research if they do not have the opportunity to refuse consent to research on their samples or, at least, to opt out.

> I am concerned that what will happen is that it may end up practically being: if you want a diagnosis done then you have to donate your sample for research purposes. And people need to be able to opt out of that freely and I am worried that it would become 'sign here and here, and come into the lab' etc. So first of all…—as is always the case with informed consent—it's not just a question of what it says but what the informed consent process is … you need to make sure that people understand that this is a separate issue from getting diagnostic testing done (#36 US, medicine, genetics, government, R/A/S).

Arguments in favor of new informed consent and against general consent   According to respondents favoring new informed consent, using samples taken in a clinical context is ethically more problematic than using samples taken in a primary research context because ill patients are more vulnerable than healthy research participants and therefore need greater protection:

> I know that scenario D is similar to scenario A, but here the samples are collected for diagnosis. So my answer is … any time I want to use them, I need to get [new informed] consent. In scenario A the participant is healthy and voluntarily donating the blood; but in this case, it is more difficult for the donor to say no. So we need a policy that says clearly 'only diagnosis,' and if you want to use them in research you have to go through the usual procedure (#13b IS, No. Am./Asia, law, university, A/R).

The reasons why the option of new informed consent was chosen more often in scenario D than in scenario A seemed linked to the fact that the term "medical research" used for general consent was perceived to be too broad.

> Any consent is better than none but I would … make it a strong consent standard [preferably new informed consent]. And the reason is directly correlated to the huge openness of what the research might be. So I could imagine somebody objecting to the very research itself that is being done whether it's on cloning or if they think it has any relationship that they would find offensive. If they think it is research that is a bad use of public money—we are spending too much on prostate cancer and I want to help with whatever [else]—then they might not want to participate in it (#32 IS, Europe, medicine, university, S/A).

Some interviewees from indigenous groups referred to cultural differences concerning the meaning of biological material for the individual. Individual consent to each use of the samples is necessary to avoid violating the beliefs of patients in societies where biological material is of particular importance for the individual.

> [Research is acceptable] provided that they [hospital patients] consent for each use. It is important that the person knows exactly what they are going to do with that sample, what they are planning to obtain, and in which moment it is going to be used for another purpose … If it is enough clarity and there is feedback from the [patient] so that we are

sure she clearly understood what is going to happen, then it is possible to carry on the research so that we don't violate their way of thinking and culture in relation to what I am doing [as a researcher] (#86 IS, So. Am., medicine, university, A).

Arguments in favor of general consent and against new informed consent   The term "medical research" used in the consent statement in scenario D elicited not only comments from respondents opposed to general consent, but also from those in favor. It was perceived to be even broader than the explanation "research on other diseases" used in scenario A. One respondent favors this term because the broad definition "medical research" maximizes the range of beneficial research for which these samples could then be used. Furthermore he appreciates that the definition is simple and devoid of "scary" legal terms. He expressed the opinion already known from responses to scenario A that simple descriptions are preferable because research participants don't have sufficient education to deal with complex choices.

> I should always try to minimize the choice for the participant, because I am afraid of so many misconceptions in the mind of people making them refuse such participation, which of course are not correct. And the more I write to them, the more they would get scared, especially myself as a scientist when I receive a research statement with legal terms I usually start shaking. So just make it very simple ... I am aware that 'medical research' is open to diagnosis and treatment. Future generations would benefit from it, other affected people would get some benefit. For this reason my acceptance is once for all (#93 IS, Europe/Middle East, biology, genetics, biobank, S).

Another respondent, although in favor of general consent, prefers to specify some limits and proposes to add "that the research is done for no profit" (#57 IS, No. Am./ Europe, medicine, university, S).

Similar to respondents against general consent, those in favor see the use of samples for research as being fundamentally different from any use for a clinical purpose. However, they think that general consent to research is sufficient, as long as it is explicit and not construed.

> I think there needs to be permission for research. A clinical sample is not automatically a research sample. And I think 'I agree to my stored specimen being used for medical research provided that I have agreed to the general category of study' is what I would like it to be. Not each exact use. [New informed consent] is too restrictive and [presumed consent] is not specific enough (#07 US, social sciences, bioethics, university, O).

Respondents against new consent pointed out that in scenario D recontacting for new consent is even harder and more expensive than in scenario A:

> The problem is—once they leave the hospital—[it's harder] to find them and to get their approval! (#57 IS, No. Am./Europe, medicine, university, S).

> [New informed consent in scenario D] does not make sense in the [my Asian country] context ... it is very, very expensive (#78 IS, Asia, life sciences, government, S/A).

Moreover, recontacting of patients after hospitalization is considered particularly complicated, potentially harmful, and not in line with many patients' wishes.

> [New informed consent] becomes really problematic, and the biggest issue is what counts as each use? So for example, I had my surgery five weeks ago, and I imagine they are going to do tons of research with my biological material, with the information they have collected about me. But what counts as one study versus another study? Is it every single paper? It is hard for me to imagine how to implement this. I certainly don't want to get [disturbed all the time] by letters from [my university hospital], if every three weeks somebody is looking at it in one way, today they're looking at characteristics between these two factors and tomorrow somebody is going to look at another factor …that's going to drive me crazy. I would rather say: use my sample for research purposes also (#59 US, medicine, bioethics, government, R/A/S).

In addition, the benefit for public health of being able to use a maximum of samples without "many losses with recontacting" (#34 IS, Europe, medicine, genetics, biobank, S) was judged particularly high in scenario D:

> My wish would be to see generic consent, option A being the preferred one because I think that's the one that more likely leads to significant health gains (#82 IS, Europe, natural science, A/R).

> [General consent] is essential … If you had an emerging infectious disease and you have a collection from a hospital from patients, it is the mission of medical doctors to search for infective agents in those samples. That is important. By bringing the issue of infectious disease, it makes it easier because people understand it (#67 IS, Asia, genetics, government, biobank, R/A).

Several respondents said they agree with general consent in scenario D.1 only if specific safeguards are added. Such proposed safeguards were irreversible anonymization of the samples (#72 IS, Europe/Africa and Europe [nationality: Europe], medicine, S/A) or limited use of samples for specified future projects only, together with IRB approval (#36 US, medicine, genetics, government, R/A/S).

Arguments in favor of general consent and against presumed consent One respondent argued that adequate protections for research participants who provided clinical samples are the most important condition for allowing future research use. He questioned the right to opt out because it might imply negative consequences for minorities who have a disproportionate tendency to refuse participation in research:

> [General consent] to me is the best because it is basically giving people notice that their samples can be used for research … I think the question is whether or not they should have the ability to refuse that and the thing is that from a purely clinical perspective you don't want to use samples for clinical studies, as long as adequate protections are in place. If you get people who are opting out, for instance, let's say African Americans are less trusting on the system, say African Americans would be more likely so say that they don't want their samples to be used for research, at all. And then some doctor wants to come along and say: well let's see if there is something going on in colon cancer and he's going to pull 200 samples off the shelves of colon cancers, but as it turns out, because of that consent

process, he has got very few African American samples, he's under-representing African Americans, is that good? I don't know, I think that's bad for the community and that's bad for science (#10 US, humanities, law, university, R/A).

The need for collective consent   Several respondents from developing countries or indigenous groups brought up the issue of group consent in scenario D.[2] These interviewees would prefer—in addition to individual general or new informed consent—to obtain group consent for the research use of biological samples taken for clinical purposes.

> [W]e are talking about the heritage of the indigenous community … it is important that we must have the acceptance of the *council of ancestors* because we take individual informed consent from each person (#86 IS, So. Am., medicine, university, A).

> I was thinking: how well do the indigenous groups understand this, unless there is someone to protect their rights. We have tribal belt commissioners who are supposed to protect their rights. So if there is a similar group or mechanism in place and if they can talk to them and ensure what they want their samples should be used for. Then it [general consent] is ok (#42 IS, Asia and No. Am./Asia, medicine, bioethics, government, A).

Even in the US, a respondent asked for an equivalent of society consent in form of "broader discussion with the public":

> The surgical consent forms in the US for a long time have said that samples can be used for research. I don't think that's meaningful. What I do think is that there really does have to be some oversight about how these samples are used. I do think that there needs to be a broader discussion with the public about what happens when you come to the hospital and when you have a surgical procedure. I think [general consent] is unacceptable, because there is not enough there to be meaningful (#92 US, law, medicine, genetics, university, R/A).

2.2. Consent and anonymization in scenario D.1

A majority of all respondents stated under D.1 that at least some form of consent must be obtained, even if samples were anonymized and irreversibly unlinked to the participants. Only about one third of the interviewees from the US and Europe and one fourth of all other respondents supported the opposite idea that research without consent is ethically appropriate if samples collected with prior consent to diagnostic testing are irreversibly anonymized. Almost half of those who had used or stored samples, compared to less than one fourth of respondents who had not been involved directly with samples said that consent is not necessary if samples are anonymized.

Arguments in favor of research with anonymized samples without any form of consent   Respondents argued that consent is not needed because if samples are "irreversibly anonymized, the individuals from whom they originated have no longer any interest in them" (#82 IS, Europe, natural science, A/R). The benefit

---

2 See also the following chapter on collective consent (7).

for public health is said to justify research with anonymous samples, especially if risks to confidentiality and to privacy are considered to have been eliminated with anonymization.

Some respondents answered that it is not sufficient to anonymize samples and associated information irreversibly. Only after approval by an ethics review board would it be appropriate to use these samples and information in research without consent of the research participants.

One respondent from the US explained that anonymization might not agree with patients' wishes. This policy might be acceptable, in need of justification. To assume that "so long as we anonymize that we don't have any problems I think that's a mistake" (#92 US, law, medicine, genetics, university, R/A).

> I think that what we have to recognize is the decision to unlink them and not to make the effort to reach out and notify people about research is an ethical decision. So long as we understand that we've had the opportunity to talk with them about it, we've chosen not to exercise it and we have chosen instead to anonymize their samples, even if they might not want this to go forward. I think so long as we are clear about that, then we can decide whether that is a good or a bad policy decision (#92 US, law, medicine, genetics, university, R/A).

One respondent pointed out that the need for consent varies according to the "risk benefit ratio … if the benefits are so high and the risks so low, then it may be ok" (#96 US, genetics, medicine, social sciences, university hospital, R/A/S). For another respondent, it might be justified to use samples without consent, if it is "completely impractical" to obtain consent to research in the clinical context.

> I don't think that by virtue of providing medical care, a hospital has the right to use samples for research without getting consent. So I don't think that's acceptable. I might be persuaded to change my mind if it's completely impractical to do it (#45 US, bioethics, philosophy, government, O).

Arguments against research with anonymized samples without any form of consent   Several respondents who in principle approve of the use of anonymized samples without consent insisted that researchers have to obtain informed consent from research participants before proceeding to irreversible anonymization of their samples.

> [O]ne should say in the consent that 'I am collecting your sample, anonymizing it, and do whatever I want to do with it.' (#56 IS, Europe/Europe and Africa, medicine, life sciences, government, hospital, S/A/R).

Whether this more relaxed standard of consent to future uses of samples in anonymized form should be called informed consent remains controversial:

> For me the question is 'what do the persons consent to when they give the sample?' If there are guarantees that the samples will be completely anonymized and that they are giving their consent for all uses in future, then we have consent to do it. It is ok but I

would not say that that is informed consent anymore (#46 IS, No. Am., law, social science, NGO, A).

Interviewees argued that, if not consent, so at least notification of hospital patients is ethically required to comply with the principle of respect for autonomy.

It's just respect (#64 US, medicine, bioethics, university hospital, N).

In contrast to respondents who think that consent is not required for the use of anonymized samples, respondents holding the opposite view consider that the interests of research participants are not limited to the protection of confidentiality, but include an interest to know what happens to their biological materials.

Confidentiality is not the only issue. There are other issues. This sample did come from a human being that has values and preferences and interests in certain things. So I am not saying that it is ethically unacceptable to use these samples but you should ethically notify people about what is being done to their samples, send them a newsletter, update them on the outcomes of the research. It is not like if you took care of confidentiality you took care of everything. There are other ethical considerations (#39 US, Europe, philosophy, bioethics, government/university, R/A).

Respondents mentioned in particular that it is in the interest of research participants to keep samples linked in case they would like to benefit from the return of research results.

At a stage where it [the sample] becomes anonymized, it removes the potential for benefit back to the participant, so they need to be made aware of that too. So for some people that would be ok, but other would say 'no, I don't want my DNA to be used in that way' (#14 IS, Oceania, political science, indigenous IP, university A/R).

Other respondents argued that to respect patients' autonomy, one time consent to anonymization is not sufficient because the patient is not able to foresee to what he or she agrees in a meaningful sense.

If the patient consents that the sample is [anonymized] the problem is when the patient initially gives his consent under certain conditions and then the conditions change (#86 IS, ind group, So. Am.).

For some, respect for autonomy entails that anonymized samples should not be used for research if this is against patients' wishes. Respondents favoring consent referred to empirical data showing that "people don't want it [...this is] not my own data but colleagues of mine have done studies about that. And people do want to know what their samples are used for" (#07 US, social sciences, bioethics, university, O). This view is shared by another respondent from the US who referred to his personal preferences:

I don't think people's material should be used in anyway, unless they know that it could be used that way. And although we say that it is completely anonymized, that may not be. So I always thought, when I gave my blood that it was for diagnostic testing, to get back to my doctor for my care and I believe that not just I but anybody whose materials are going to be used otherwise should know how it's going to be used. I don't know how that can be done without individual contact. And I think that at the very beginning I ought to be [able to] say that I am giving you this sample for diagnostic purposes and want to opt out of any research, that choice should be given (#09 US philosopher, catholic theology, university, N).

As shown by the previous quotation, respondents from the US think that individual consent is necessary because they are convinced that anonymization might not sufficiently protect the identity of research participants. Anonymizing would still mean that there is a requirement of consent "because you cannot absolutely anonymize anybody's sample. Unless what you are saying is: the sample is only a segment of DNA, which isn't long enough to be unique" (#80 US, genetics, bioethics, philosophy, theology, university, R/A/S). Another respondent argues that group consent instead of individual consent is required because he is concerned about the fact that although individual samples are anonymized, research results might still cause harm to groups.

It depends on how identifiable the population is … if it is just individuals of various ethnic groups, then in my mind, there would be no need to contact the individual. If it is a socially identifiable group, then I think there needs to be some contact with the leaders if there are leaders. The group could be stigmatized or discriminated against (#70 US, genetics, social science, bioethics, university, R/A/S).

One respondent from Europe mentioned "ethics approval at a collective level" as sufficient substitute for group consent in Western societies without traditional "leaders."

Probably [consent is] not [required]: if fully anonymized with ethics approval at a collective level, you might get some estimates of prevalence of diseases (#53 IS, Europe, medicine, public health, Int. Org., R/A/S).

The requirement of consent for research use applies if the specimens are irreversibly anonymized and unlinked to the patient "because even though the privacy interest is not at stake, the autonomy interest is" (#41 US, bioethics, law, university hospital, R/A). Interviewees are aware of cultural differences as regards attitudes towards human biological materials. As explained by a respondent from the US, these differences have to be respected, in line with the principle of respect for autonomy.

Some people are opposed to some kinds of research, suppose in the US, the sample is going to be used on, let's say abortion research. A lot of people would be bent out of shape about that. You have a right to decide what is done with your samples and some communities—in some Indian tribes it's unacceptable to have samples retained. They have to be buried with you when you die. So it doesn't really matter whether it is anonymous or not (#41 US, bioethics, law, university hospital, R/A).

Interviewees from Africa and Oceania indeed expressed specific attitudes towards samples. They insisted on the importance of ongoing relationships with research participants in their cultural context, which is incompatible with irreversible anonymization (#77 IS, Africa, social sciences, NGO, A) and described how samples are representative of the identity of a person in terms of his or her genetic heritage and genealogy.

> Even though the samples might be anonymized, they still come from people; they still belong to the heritage, genealogy of this person, so they still need to be asked (#14 IS, Oceania, political science, indigenous IP, university A/R).

Consistently, respondents from indigenous groups as well as from countries outside the Western European and North American culture indicated that they want to be "told that the sample will be used" even if there is no risk "that my privacy will be disclosed" (#13c IS, No. Am./Asia, law, university, A/R).

> If my sample was going to be used in any form, even anonymized, I want to say 'yes' to it, or consent to it (#37 IS, Oceania, social sciences, bioethics, university, A).

Other responses indicate that the desire for transparency and the wish to consent might be influenced by the experience of previous research or the fear of future abuses. Interviewees said that these fears provide sufficient reasons for policy makers to choose consent because the benefit of science depends on the trustworthiness of research institutions.

> I think you could do that [use anonymized samples without consent]. I think there could be unpleasant PR implications. I am aware of some cases in Europe and it was autopsy samples … you could have easily told families, but it wasn't done and people were very upset to find out about this. So I think a one-time consent process goes a long way towards transparency and perceptions of trustworthiness. So I think it serves several purposes (#26 US, medicine, social sciences, government, R/A).

2.3. Research using existing collections: is it justified to use samples without re-consent?

Under D.2 the problem of existing collections was addressed. Respondents were faced with a policy providing that:

> Where a research project involves linking a stored specimen to a patient who has not provided the necessary consent for the research, an investigator should recontact the patient to meet the consent requirements. If, however, a patient cannot be recontacted through reasonable efforts, the investigator may apply to the Ethics Review Board to use the specimen. The investigator shall show that reasonable efforts have been made and that the aims of the research cannot be achieved without using such specimens.

Respondents disagreed on whether a policy requiring that participants be recontacted through reasonable efforts is adequate. Between one half and two thirds of all respondents agree with this policy. Many of the remaining interviewees think that recontacting should be proscribed because it is intrusive and risky for the privacy of research participants. Respondents were also divided as regards the second part of the policy, i.e. the permission to use the samples. Some of them considered that re-consent is not necessary and samples could be used after approval of an ethics committee. Others, on the contrary, are convinced that existing collections should not be used at all. In summary, disagreement exists between at least four groups of respondents:

(1) Respondents who think that recontacting is required. Without re-consent, samples should not be used at all;

(2) Respondents who agree to most aspects of the policy proposed in scenario D.2. They favor reasonable efforts of recontacting and allow for research with ethics committee approval if these efforts fail. The definition of "reasonable" varies widely between not even trying while showing that recontacting would be disproportionately costly, all the way to demanding public notification as well as phone calls and letters to all research participants;

(3) Respondents who find recontacting wrong (intrusion, too cumbersome) and who favor proceeding with research after ethics committee approval and/or anonymization without any attempt at recontacting.

(4) Respondents who find recontacting wrong and don't allow any use of samples from existing collections outside the range of the consent provided by the individuals at the time when samples were taken.

Numerous respondents from the groups (2) and (3) were from Europe and North and South America, whereas the positions described under (1) and (4) were defended by a substantial number of respondents from developing countries and indigenous groups (as well as some respondents from the US and Europe) and motivated by fear of abuse and past problems of abuse. In the following we will examine in more details the arguments used by respondents from the four groups.

The arguments of group (1): without re-consent, samples should not be used

Among the critics of the policy in D2 were first of all those defending classical informed consent. They noted that researchers may use samples that were collected without consent only after they have been able to secure "actual" or "direct" re-consent.

> If ethically informed consent is required for the use of samples and information and if I can't achieve informed consent then we can't use the samples. If we can do that [obtain re-consent] that is fine. But it goes on saying that if we cannot ... then the ethics board may give permission, so that we can go on using the samples ... that is not the patient's informed consent ... If samples were obtained in the past without consent, we can't use them (#71 US, Europe, law, philosophy, bioethics, university, N).

The argument that it is unethical to use samples without classical individual informed consent was invoked by respondents from different parts of the world.

> [The] ethical part of it [is]: as long as I did not get the direct consent of the patient, I cannot use the samples (#93 IS, Europe/Middle East, biology, genetics, biobank, S).

If there is any way that is ethical to use them, we should try not to destroy them. If a patient could be recontacted, then you should do that. ... If we cannot recontact the patients we should consider destroying them (#13c IS, No. Am./Asia, law, university, A/R).

Several respondents from developing countries and indigenous groups further resorted to the principle of non-maleficence as justification for their opinion. They referred in particular to abuses in the past and the potential for future abuses.

There is very good research but also a lot of very sloppy research, and the potential for abusiveness is quite significant (#14 IS, Oceania, political science, indigenous IP, university A/R).

One respondent from Africa indicated that in order to reestablish trust, sample use without re-consent should be strictly prohibited. He thinks if consent cannot be obtained, it is preferable to "start again" and to establish new collections to carry out research.

We are having a lot of trouble now in [my country] because babies' hearts were taken without permission. And that was all found out, and the hospital has been sued now because of taking of specimens and using them. They might have taken them 50 years ago. But when it came out, the whole case erupted and they have now been sued for obtaining body organs and things. It has just been the most dreadful thing happening in [my country] with body parts being taken before there was need to take consent or something ... Whatever happens, you need permission. You can't use [them ...]. I would say you have to start again because it just goes against everything else we said before and it brings up to me [all problems with] body parts, cells, or whatever. I think that you should start again. This is my personal opinion. There are all different ways to look at things and I can't speak for [the entire indigenous group.] I can give you a flavor of what is happening. I just don't want to be quoted as 'all [indigenous people in my country] do that'. It is difficult (#37 IS, Oceania, social sciences, bioethics, university, A).

Another respondent working in Africa gave an example involving serious harm and explained that in his opinion consent is the most appropriate way to protect against abuse. He thinks that it is "better to be careful" and not to use samples without new consent. Even if researchers anonymize samples irreversibly, research participants should give "permission for a very specific use" (#56 IS, Europe/Europe and Africa, medicine, life sciences, government, hospital, S/A/R).

If a participant cannot be re-contacted [... samples cannot be used because ...] it is dangerous. Some of these arrangements would favor the kind of biomedical research I am involved in, but I see all the risks that could materialize if something goes wrong. It is better to be careful, because, if something goes wrong, then you have to stop everything ... Consenting protects from linking demographic data to samples, which is very important here, because if one could access samples to research genetic susceptibility to HIV, make a list of all HIV positive participants, and blackmail them, it would be dramatic. But these are obvious issues, protecting demographic information. But also the context is important (#56 IS, Europe/Europe and Africa, medicine, life sciences, government, hospital, S/A/R).

Other respondents mentioned respect for research participants as the reason why "if the person cannot be recontacted, [the sample] should not be used." The ethics committee should not be allowed to authorize the use of samples in this situation because "I think we should respect them [research participants]" (#21 IS, Europe/ Asia, philosophy, university, A).

Respondents are aware that existing collections of samples might be valuable and that destroying them because re-consent cannot be obtained hampers research that is beneficial for public health. However, they defend that respect for autonomy outweighs all concerns about costs and losses to public health. This balancing of ethical principles is based also on the assumption that research is different from public health emergencies that might sometimes justify interference with autonomy rights and also that, in the case of research, alternatives always exist to carry out the research using samples obtained with appropriate consent.

> Even where you have samples from 30 years ago, and you have all the follow up and to set up a similar study would take another 30 years; or if [samples are unique because they have been taken during a historical period of] exposure to an environmental hazard ... possibly, you can always come up with some examples. National emergency etc., abrogating people's confidentiality for public health ... but that's different than research. And it is hard to imagine examples that [research] questions can't be answered without getting the consent (#04 US, medicine, genetics, university hospital, R/S).

Interviewees disagreed about the question whether researchers should be allowed to use samples from deceased donors. Some of those who require re-consent also acknowledged that this conclusion implies that samples of participants who have died before investigators propose to use them can *never* be used without their consent. Interestingly, this view was expressed by a respondent from the US although federal law in this country stipulates that deceased persons are no longer considered research subjects protected by the same strict regulations as persons alive.

> They should try to make every effort to recontact. But if you go to the next step—they can't contact them— to allow the Ethics Review Board to sort of consent for them, I think that's not ethical. They cannot use the samples. [Even if the participant is known not to be alive] they can't use them (#46 IS, No. Am., law, social science, NGO, A).

By contrast, others said that samples might be used if the donor is dead, as long as they are irreversibly anonymized.

> [In the case] of dead participants, if they [samples] are anonymized, you can use them freely without consent. If the person is alive, then it is probably a question of what is a 'reasonable' effort. If you don't have informed consent [from people who are alive] ... you cannot use them. I don't think it is fair [to have the Ethics Committee authorizing you]. If informed consent cannot be obtained, it would be only up to whether the person is alive or not. This is one part where you have to regulate differently [than in scenario D.2]. If the person is alive you have to get informed consent, otherwise you cannot use them (#24 IS, Europe, medicine, consultancy, A/R).

The arguments of group (2): in favor of the policy described in scenario D.2

When confronted with the question of whether researchers may be authorized—if re-consent cannot be obtained—to use samples if they show that the research goals cannot be achieved without using such specimens, the majority of respondents thought that such a policy would be appropriate. Researchers must collect new samples unless they prove that there is something special about those specimens or at least such a requirement would provide the ethics board with the appropriate information to make a decision.

In contrast to respondents from group (3), interviewees in this group agreed that reasonable efforts should be made to recontact sample donors. Their attitude seems to be based on the perception that recontacting is feasible and without significant risks or prohibitive inconvenience and should therefore be tried first. This judgement is clearly influenced by regional specificities. Respondents from the US reminded that the new consent, although not completely impossible, "is a big problem in the US where 50% of the population moves every year" (#33 US, bioethics, medicine, government, A). By contrast, a respondent from Europe insisted that in scenario D.2 "if the person is alive, [recontacting is preferable] perhaps because I live in a small country" (#48 IS, No. Am./Europe, philosophy, university, R/A).

Respondents from this group differ from respondents from groups (1) and (4) with respect to their trust in ethics committees. Although not all respondents from group (2) agree about the criteria that justify the use of existing collections without consent (see below), almost all of them wish to leave the final decision to an ethics committee (e.g. #31 IS, Europe; #65 IS So. Am.; #33 US; #48 IS, No. Am./ Europe[3]). Trust in ethics committees is found across different regions. According to a respondent from Asia, if recontacting is difficult, the ethics committee may decide, as long as it provides justification for the use of the samples.

> We are also talking about that in [my Asian country]. But we also have a situation where in some of the institutions where you have a lot of mobile population coming in as a patient group, recontacting is very difficult. The ethics board will be able to decide accordingly ... they have to justify ... why they want to use these samples (#42 IS, Asia and No. Am./ Asia, medicine, bioethics, government, A).

For those who think that the donors are the owners of their samples, this is the reason why first recontacting is obligatory ("whenever possible you should recontact the patients because I believe that the person is the owner of the genetic materials," #65 IS, So. Am., life sciences, genetics, university, A) and if this fails the ethics committee becomes the custodian of the samples. It "should analyze each case from the scientific [point of view] if the material can be used or not" (#65 IS So. Am., life sciences, genetics, university, A).

---

3 #31 IS, Europe, medicine, genetics, university, biobank, S, #65 IS So. Am., life sciences, genetics, university, A, #33 US, bioethics, medicine, government, A, #48 IS, No. Am./Europe, philosophy, university, R/A.

Never use any material if you have not [ethical approval] and you have nobody else to permit [the use.]. In this case, the ethics committee would have the custodianship of the samples (#65 IS).

Some consider the approval of the ethics committee equivalent to a form of collective consent:

> I think you should try to recontact everyone, get informed consent. For those you cannot reach, which might be the majority, you get a collective approval by the Ethics Committee. That I would feel is feasible (#57 IS, No. Am./Europe, medicine, university, S).

Several respondents from different parts of the world mentioned the public health benefit as a reason to justify the use of samples if new consent is too difficult. Let ethics committees decide for existing collections, "that's what we do actually because of the public interest." (#61 IS[4] and similarly #62 IS[5]). These respondents consider the option to carry out research with the permission of an ethics board ethically acceptable because it is viewed as the adequate balancing between autonomy rights and scientific progress. They believe that this type of balancing would meet public acceptance and receive support even from the sample donors.

> What is often phrased as a conflict between the rights and interests of researchers and research subjects I don't think is really a 'one against the other'. The public has a very strong interest in research going forward because they are going to reap the benefits of new cures and treatments and researchers have a very strong interest in honoring the wishes and privacy and interest of the research subjects so that they will participate, that they will politically support funding and so on. And so it's here where you've got archived samples and in other areas: it's trying to strike a balance between legitimate research done in an ethical manner and protecting the patients' and samples' rights. And here where you cannot reasonably contact them to get consent, and the research cannot go forward without it, then I think it is legitimate for the ethics board to consider (not necessarily be forced to but consider) waiving the consent requirement (#41 US, bioethics, law, university hospital, R/A).

The crucial difference with respondents from group (1) is that group (2) accepts the idea that autonomy rights are limited. New consent shows respect for research participants, but autonomy is sufficiently respected if an effort is made even if this effort does not permit to get in contact with the sample donor.

> Yes, it [re-consent] is a symbol of respect and it is an indication that new information has arisen, new ideas have been developed, new kinds of research are being proposed, so e.g. you might have given a blood sample for a national health survey but now people would like to work on your DNA. I think that is new enough and different enough to require recontacting. ... I still think you make an effort to recontact people; that is a token of your respect for them, even if you know that a lot of people have moved or died ...you never should do without making the effort. But you don't have to have 100% success in recontacting (#54 US, bioethics, protestant theology, philosophy, university, R/A).

---

4 Africa and Europe/Africa, life sciences, university, R/S/A.
5 Europe and No.Am/Europe, medicine, law, university, R.

Others emphasized that it is important to find a compromise for existing collections in situations as depicted in scenario D, where the law has changed towards more strict consent requirements after the samples had been obtained.

> It [strict informed consent] is part of the overall idea that people should have some say on whether or not their samples are used ... things change and changing circumstances could change someone's decision, legitimately. So here the law has changed. So now the question is what do you do for all those people before the law, well what you can do is this [the policy described in D.2], this is fine (#80 US, genetics, bioethics, philosophy, theology, university, R/A/S).

The idea that public interest justifies sample use without consent under certain conditions includes, besides public health benefits of research, considerations related to the value of the samples, efficiency and inconvenience, as well as the costs related to the remaining alternative, namely collecting new samples with appropriate consent.

> If it is easier for me to collect new samples, I would go and get new samples. Why should I use these old samples? If it is really convenient to use these collected samples because it saves me a lot of work [then I might want to be allowed to do it] but of course you should explain why you want to use them (#57 IS, No. Am./Europe, medicine, university, S).

A respondent from the US explained that "you don't require more than reasonable effort because the information and material that's already been gathered constitute a significant good by itself and you don't want to just give up on that" (#02 US, philosophy, catholic theology, university, N) and another gave an example of such valuable research on existing collections.

> I would approve it [the policy in D.2]. I have been involved with a couple of these look-backs practically speaking. We've had one to try and find out who might have been exposed to hepatitis C through blood transfusion. They are impossible, they don't work. Especially, the bigger the country, the harder it is to do it. Chasing blood transfusion records from 15 years ago in the US turned out to be a completely impractical thing to do, and very expensive. So do I approve it? Yes ... if the aim of the research was so important and you had an ethics committee that said there is no other way to do it. I might, depending on circumstances, allow that (#98 US, philosophy, medicine, genetics, medicine, university, R/A/S).

In addition, sample use without consent is not justified only by public benefit, but also by the principle of non-maleficence concerning future patients. Taking new samples means risks for more patients and these risks for future patients have to be balanced against the risks for the donors of the existing samples:

> This issue of whether the research can be done in other ways should always be contemplated in an IRB [research ethics committee] application. If there is another way to obtain the same data that is more ethically feasible, of course it should be used. But this is often not possible or very difficult ... I find that [collecting new samples] highly ethically questionable because of the substantial medical risk involved in obtaining specimens. The issue one should consider is whether there is an ethically more favorable option to use

all these supplies, [the option of using existing samples without re-consent might] have problems with autonomy but in other instances it may be instead favorable because of not putting the patients at medical risk (#32 IS, Europe, medicine, university, S/A).

Some respondents referred to another interesting aspect related to benefits and harm. They argued that it is justified to use samples for which recontacting fails because in this case the risks are very low. The underlying assumption is that risks for sample donors consist in genetic discrimination and stigmatization if research results reveal predispositions to disease or undesired other traits: "What's the harm? The harm in this case is that you might blurt something to the individual ... if you can't contact them, you can't do that" (#33 US, bioethics, medicine, government, A).

However, even if risks are minimized because it is not possible to recontact an individual, some respondents would still require higher standards as proposed in scenario D.2. They proposed additional strict protections for confidentiality, including anonymization.

> I think for the reasons that I just mentioned, i.e., transparency and trust, it is a good idea to inform people and give them an opportunity to consent at some point in time to this but I also think that since the primary risk is that of contact and providing information, if you can't contact the person, it is also an indication to me that that risk is minimized in many ways so I think you can sort of ... well, I might consider anonymizing the samples or want really clear confidentiality protections and a commitment not to recontact anyone with results if they haven't given consent (#26 US, medicine, social sciences, government, R/A).

This respondent from the US is not the only one who basically agreed with the policy in scenario D.2, while proposing some modifications. We will first summarize the arguments of those who favor less strict standards than described in scenario D.2 and then report on proposed higher standards. The answers of respondents summarized above have already shown that for several among them, the reason why samples might be used without consent should not be restricted to the condition that the aims of the research cannot be achieved without using the existing collection. Instead, a substantial number of respondents indicated that the ethics committee or institutional review board (IRB) should balance benefits expected to result from the study with harm to research participants and patients.

> It depends on the particular study. Applying to the IRB is ok, and the IRB will look at the particular study before they decide whether consent has to be obtained or not. I would look at a particular scientific value or a particular association that may come out of this or how the patients might be possibly be affected by this study (#19 IS, Europe/Middle East, genetics, university, S).

As a consequence of giving priority to risk-benefit evaluations, IRB decisions are expected to vary from case to case.

> I guess I would argue that it might be case specific. In some cases I would say yes. In other cases I would wonder if either the research wasn't all that important, or the risk of having the specimens used without consent was so great that it shouldn't be done (#43 US, bioethics, medicine, philosophy, government, O).

Others seemed to consider the risk related to studies on existing samples usually low and required only evidence of scientific merit of the study. Whether research without consent is acceptable "would depend totally on whether we view the scientific aims as meritorious" (#01 US, medicine, bioethics, university, N).

Several respondents prefer not to define any justifications in advance. Investigators "should be able to [satisfy the requirement] by making a statement" (#30 IS, Europe, medicine, bioethics, government, R/A/S). This statement could invoke different reasons. "There might be something unique about these samples that it would be difficult or impossible to replicate. It might be a financial question or there might be a whole range of reasons why you can't use other samples, but you must show these and have thought about [them]" (#82 IS, Europe, natural science, A/R).

Finally, for one respondent from Asia the only condition to justify the use of the existing collection is that recontacting failed: "they just need to show that they made an effort to contact them." (#74 IS, Asia, genetics, hospital, S).

Respondents who are not entirely satisfied with the policy, feeling that the standard for use without consent should be *even higher*, proposed for example group consent. One respondent from Africa believes that otherwise the requirement of "reasonable effort" is an alibi for abuse.

> I don't agree [with the policy in D.2]. It happens all the time that the researchers can say that they have put a reasonable effort when they haven't done anything. I go back to my basic point, informed consent is collective. If you cannot contact them individually, the family should be contacted, and then there is the possibility to form an association through which participants are kept informed about the research. So that takes care of the question of making the effort to contact the donors. A reasonable effort is not enough. If you can't get the family to consent ... the group itself must have a procedure in place ... let's say 3 years from now, this is who should be contacted. It [prevents the] possibility that researchers say that you did everything but you could not contact (#77 IS, Africa, social sciences, NGO, A).

The disagreement about the standards that the ethics committee should use extends to the question of what reasonable efforts are. Interviewees pointed out that the formulation of the policy is not satisfactory, invoking the indeterminate or vague character of the word "reasonable."

> It could be religiously applied or it can be quite lazily applied. It needs to be spelled out. A reasonable effort might be that you make one phone call, or you send one letter and the person has moved. So I would need to see 'reasonable' defined in some way (#14 IS, Oceania, political science, indigenous IP, university A/R).

Respondents described a range of means they would consider to be reasonable efforts in scenario D.2. For some respondents, the essential part is that efforts are really made, instead of simply assuming that these efforts will be unsuccessful.

> I would put a high threshold. I certainly don't want them to just come in and declare that it is infeasible. I want them to make the effort (#54 US, bioethics, protestant theology, philosophy, university, R/A).

One respondent explains that what is reasonable varies according to the circumstances.

> [What would be reasonable efforts] that's a very good question. That's something we debate all the time, and that, too, depends on the circumstances. For example, if I wanted to contact everybody at the university, we send e-mails all the time, no one sends out letters. That's considered reasonable effort to contact everybody. Does that mean you could do that for a rural community in Montana, I don't think so. There you might have to send out letters, other places you might need to make phone calls (#80 US, genetics, bioethics, philosophy, theology, university, R/A/S).

Reasonable efforts might comprise advertisements in the newspaper.

> Reasonable efforts [could be] telephone, mailing a letter...You could alternatively put an advertisement in the newspaper saying: we are going to do this. If you have ever been a patient at that hospital, call us. I don't have a set idea about what is reasonable. One would have to talk it through. If it's two million dollars to notify people by mailing, maybe [it's still reasonable]. I don't know. I don't have a pre-conceived idea about what is reasonable. I do think that there are likely to be thousands more people coming to the hospital from whom you can collect samples and sometimes we get carried away ... we should make very substantial efforts to let people know (#28 US, bioethics, medicine, government, N).

Several respondents think that costs shouldn't matter.

> Working over a month or so, letter, telegrams, telephones, and that should all be recorded. We should have that information down, so that it can be presented to the ethics committee to say: this is what I did to contact the patient ... I don't care how much it costs...because we have to have their consent, any individual whose material is being used by science has to consent. That's just a universal ethical norm (#63 US, bioethics, medicine, catholic theology, university, A).

Among the safeguards that should be added to make the policy in scenario D.2 acceptable were, according to the views of respondents from the IS and US, minimal risk, anonymization (#25 US, philosophy, government, N) or "confidentiality provisions" (#26 US, medicine, social sciences, government, R/A), exclusion of controversial and worrisome research and the fact that "the Ethics Review Board must absolutely contain meaningful community representation" (#36 US, medicine, genetics, government, R/A/S).

The need for high and possibly costly standards was justified as ultimately beneficial for research, because if a strict policy is in place, the community is less likely to lose trust than would be the case with permissive policies:

> It is still desirable to try to contact the people. I could understand some situations but I don't think it is good for research in general. When the community becomes aware that 'oh, they have decided to use anything that's available up there on the hospital for whatever they want to do' there is potential for people to lose trust (#36 US).

Whether anonymization should be required as an additional standard is subject to controversy. Some respondents are convinced that anonymization is necessary whenever a risk exists to inform participants about research results, whereas others defended the contrary:

> It also depends again on what you are going to use [the samples]. Suppose the scenario, in which you have already information that a certain gene or number of genes are associated with a certain disease, and you are providing confirmatory evidence that may be used for diagnostic purpose or for treatment, then it's maybe not necessary to anonymize them because you decide 'yes I am going to inform my patient of that' (#60 IS, So. Am, pharmacology, medicine, government, S/R).

Finally, one participant from the US proposed a unique empirical model to define when ethics committees should approve the use of samples without consent. In order to allow the conclusion that "no reasonable subject would object to their use without permission" IRBs should be able to prove that they have done an empirical study indicating that "with a p value of 0.05 you can conclude that the subjects would not object" (#38 US, bioethics, philosophy, theology, medicine, R/A).

> It doesn't mean that merely reasonable efforts will be sufficient to justify the use of such samples. What I would favor would be a requirement to make reasonable efforts to contact the patient and if the patient cannot be contacted, there must be an ethics reviews board decision that no reasonable subject would object to their [i.e. samples] use without permission. And I would add to that, it seems to me very unlikely that there is ever a case where no reasonable subject would object ... the burden of proof is on them [the IRB]. So if someone comes along and sues them because their sample was used without permission eventually the IRB would have to defend its conclusion that no reasonable subject would have objected. And since there is somebody objecting, the IRB would have to prove [showing empirical data] that that objection was unreasonable ... my notion of empirical data is quite soft. Whether they have actually gone out and interviewed people or not is not to me absolutely essential. I have been on IRBs that have made these judgments by surmising that no reasonable subjects would object. But it seems to me far better if you have got empirical data. At least the IRB is then protected. If an IRB is sued for releasing data [and samples] without subject permission and the IRB has gone and done the empirical study and finds that [with] a p value of 0.05 you can conclude that the subjects would not object. That is the only definitive defense of the IRB's behavior. If they didn't do the empirical study and they say: well we are sure people would not have objected, they are vulnerable for somebody claiming: I do object therefore your conclusion must be wrong (#38 US).

The arguments of group (3): research without re-consent

The respondents in this group are divided: some think that irreversible anonymization is required and others reject it.

In contrast to respondents from group (2), interviewees from group (3) who do not think that anonymization is required argued that "an ethics review board should have the possibility to waive the requirement for consent even if it [re-consent] has not been tried ... for example if it is a large number of samples or the samples are very old" (#32 IS Europe and similarly #79 IS So. Am.[6]) A respondent from South America described that this is current practice in his country.

> We have many biobanks that have been created only for health treatment but we are using them for research. We have this material from many years ago, like tumor banks where people have died. The researcher has to sign a document to the IRB saying that they are using this to do only this project with this objective, and not using them making a link between the material and the personal information. [They are allowed to use medical records, too], but our hospital has electronic medical records. If one wants them, we can send them without any identifiable data (#49 IS, So. Am., biology, bioethics, university, A/R).

Many of the arguments used by respondents from group (3) against re-consent resemble those employed by respondents from group (2) in order to defend research on samples without consent after failure to recontact participants, namely that in certain situations public health benefit and the burden of recontacting outweigh individual rights. But they go further arguing that for these reasons "some research can be done without consent." Hence, this respondent would "put the ethics review in a different place ... further up in the process" (#43 US, bioethics, medicine, philosophy, government, O) allowing for the possibility to waive re-consent entirely, especially if re-consent would be very costly.

> It is very difficult to recontact every patient, especially if the stored material is large [i.e., there is a large number of stored samples]. I believe that, if the investigation was made with all the ethical principles in mind, the recognition of these collected data can be made independently of contacting every patient or every person in order to have their consent (#79 IS, So. Am., medicine, university, S/A).

A proposed safeguard for this practice is discussion in society to make sure that research on samples without consent is publicly acceptable.

> Research organizations and university hospitals should pay their costs for propagating their activities in society, but recontacting patients is a very different thing. I don't think recontacting is the best way. Anonymizing them and using them without consent is one way. But a very important part is the necessity of the specimens. If the specimens are indispensable materials, we have no choice but using them. We have to raise that kind of argument in the society first. In [my country] this is a touchy issue. I am a member of the Pathological Association. Recontacting is practically impossible especially with deceased people .... Family members have different opinions and it is very tough to get family consent for a deceased person (#67 IS, Asia, genetics, government, biobank, R/A).

A substantial number of respondents would opt for research after anonymization of the samples instead of trying to obtain re-consent.

> This is secondary use. I have no objection with researchers using it without consent, as long as it is thoroughly anonymized and the ethical [issues] are clear. I think the world has gone too far towards protecting individual rights and sacrificing the collective, public

---

6 #32 IS, Europe, medicine, university, S/A, #79 IS, So. Am., medicine, university, S/A.

health dimension. I personally hope the world is going back to recognizing that we all benefit from biomedical research (#68 IS, No. Am./Europe (nationality: No. Am.), life sciences, consultancy R/A).

Anonymization is perceived to be the preferable alternative because recontacting "raises privacy issues" (#10 US, humanities, law, university, R/A).

> And when using archived samples, it is not clear to me that you need to use identifiable samples and set up a linking system. You could do something to anonymize. And I think that this is preferable to the whole consent thing. Consent exercise is problematic, recontacting the subjects is hard, it is expensive so I don't know if it is worth it (#10 US).

Others would make the choice between re-consent and anonymization taking into account the practicability of recontacting.

> I support an alternative which is that either you recontact or you have to anonymize but it depends on the difficulty and practicability. So [in the case of] 50 sample they should recontact the people. If it is 50 000 and it has been over 50 yrs, a lot of them are dead, a lot of them have moved, then I would be more inclined with anonymizing and not recontacting (#27 US, philosophy, bioethics, government, R/A).

The arguments of group (4): against re-consent—samples should only be used for purposes specified in the original consent In scenario D.2, respondents raised, more often than in scenario A, the concern that recontacting can breach confidentiality owed to the participant. For a small number of respondents, this implies that samples from existing collections should never be used.

> I would suggest to give up the idea of using them [i.e. the samples...] we need to re-identify patients and the effort to recontact them is an invasion of their privacy, and it can be more dangerous ... we do have a huge blood bank of soldiers ... the effort to contact them is not ethical at all (#13b IS, No. Am./Asia, law, university, A/R).

Two respondents in group (4) are from Asia and Europe/Africa and the third is an interviewee from the US who had not been involved with the issue of biobanking before. These participants objected that apart from being intrusive and putting the privacy of research participants at risk, re-consent is also a burden for researchers.

> It would be very burdensome for patients to re-consent to the research and for investigators to recontact the patients to meet the requirements. I don't think that's the way to go. That would be intrusive and burdensome to the researchers. I am also worried about the implication for the patients. You should collect samples prospectively ... I don't think you should use samples without people having given consent for that use. I don't think that's right (#72 IS, Europe/Africa and Europe [nationality: Europe], medicine, S/A).

The respondent from the US (#09 US philosophy, catholic theology, university, N) is against re-consent because he doubts that it is possible to obtain valuable consent after many years. He particularly fears that research participants would not understand to what they are consenting because re-consent would be done under inappropriate conditions (e.g. over the phone) and that research subjects lack the intellectual capacities to understand the scientific and ethical issues at stake.

> When they [donors] gave their material, six months ago, six years ago, they had no idea that this might be done. Most people don't follow these issues in the press. So they wouldn't be contacted face to face, they would be contacted over the phone, not by the researcher themselves but by some assistant ... I would say that the average person just does not understand the scientific [issues ...] they give this stuff 30 years ago and now they're ... being asked if they could [agree] ... So I am just afraid that this would not be conducted thoroughly enough. I would be getting calls, it would be some kind of research assistant at best, some telemarketing person who is calling up and reading a script to me and that I would not be giving true informed consent (#09 US).

Similar to respondents from group (1), these interviewees defend strictly the principles of traditional research ethics: samples should never be used without classical informed consent.

> No I would not [approve of the policy in scenario D. 2]. I think that again this is typical of scientific experimentation. What's reasonable? They should not use the material, if they got a new law. They recognize now that there has been an error in their practices and they should throw away what they got and work with the new law from now on (#09).

In addition, the risk of abuse is felt to outweigh public health benefits. Society is said to be under no obligation to give primacy to the eradication of diseases.

> I am afraid there is more abuse in the doing of this than there would be knowledge from allowing it ... we do not have an obligation to wipe out all diseases or achieve all good benefits, unless we can do so in a way that is totally not harmful and puts persons and their well-being ahead of scientific imperialism (#09 US).

2.3 Notable similarities and differences[7] between subgroups of respondents

While in the US sample, the percentage of those in favor of general consent is the same in scenarios A and D, it varies in the international sample. Europeans are more favorable to general consent in scenario D (three quarters) than in scenario A (about half), whereas the opposite was found among respondents from regions outside Europe and outside North America. Only one third of the latter preferred general consent in scenario D, compared to about half in scenario A. One might wonder whether the different results in scenario A and D are influenced by the fact that only scenario D mentions a law requiring informed consent. However, it seems unlikely that this was the reason for the differences, because one would have expected that the different framing of the two scenarios influences both the US and the international sample, rather than Europeans to react in the opposite way than respondents from

---

7 In this part, we do not include any statistical tests to prove that differences are significant. First, this is a qualitative study and many differences concern the way in which arguments are explained. When groups are compared, we only report differences that are significant ($p<0.05$, chi-square tests carried out with SPSS).

outside Europe and outside North America. If one looks more closely at the answers, the explanation seems rather to be the significant fear of abuse of respondents from developing countries and indigenous groups. These respondents noted that particular hospitals had frequently used biological material without consent. Their concern even included un-consented use of completely anonymized material. Human body parts and samples are valued differently in various cultural contexts. Those who would allow for research with clinical samples without consent if samples are anonymized are predominantly from Europe, North America and Asia. None of the respondents from Oceania and from Africa and only one from South America find research under these conditions acceptable.

The restrictive approach in certain regions outside Europe and the US was confirmed with respect to existing collections. While most respondents from Europe, the US and Asia would allow the use of samples after the approval of an ethics committee without a trial of recontacting, respondents from Africa and Oceania are inclined to prohibit research on samples if explicit consent for this purpose is impossible to obtain or is felt to be too intrusive for the sample donors.

One respondent from the US sample expressed the following opinion about differences between the US and Europe concerning the question of anonymization and consent:

> I think from the American point of view, it remains true that fears about genetic information are very much linked to penalties that may happen if your privacy is violated. So that they worry about loss of insurance and loss of job. My impression is that it is not a strong worry in Europe. In Europe it is more of an issue of just the control of information and individual human right and dignity. So [in the US] we do wind up spending a lot of time trying to keep things private. Europe in general, seems to me winds up trying to keep things consented. But I think there are illusions about cost and practicality that start to shape some of the dialogue. I have sympathy for the geneticists, who have to spend a lifetime linking the data up. If the world really wants to do this, it needs to create the equivalent of the credit card system, where you have a giant company or agency that just anonymizes etc. [and surveys the codes that link data and samples to individuals] (#98 US, philosophy, medicine, genetics, university, R/A/S).

In our study we found some support for this view. Several respondents from the US, but none from Europe answered spontaneously that the policy in scenario D.2 is only acceptable if samples are anonymized.

Scenario D.1 illustrates a different aspect of the ethics of anonymization. In this scenario, we asked in the first place for anonymization without consent. While a substantial number of respondents from the US would allow for future research without consent on anonymized samples, they would require at least consent to anonymization and this was probably the reason why they rejected anonymization as a general solution to use clinical samples for research. We also found that a lot of interviewees from Europe would allow for use without consent once samples are irreversibly anonymized.

We found another interesting difference between respondents from the US and Europe: agreement with the policy in scenario D.2 was higher in Europe (four fifths) than in the US (half). Among the possible explanations are more trust in ethics committees in Europe than in the US and less fear of adverse consequences such as loss of insurance and loss of job in Europe than in the US.

Interesting differences are found between respondents who have used or stored samples and those who have not. More sample users than non sample users are in favor of research without consent if samples are irreversibly anonymized. This would reflect their experience and be in line with their interests: researchers feel that being able to use anonymized samples without consent would facilitate research in all cases where the study does not require a link to the sample donor.

Among respondents from the US and Europe, no difference was found between sample users and non-users concerning general consent. It is the preferred option in scenario D for more than half of both users and non users. However, a remarkable difference exists among respondents from other parts of the world. In countries outside the US and outside Europe, almost two thirds of sample users favor general consent and ruled out new informed consent, compared to less than one sixth of non users. In line with this, new informed consent for each use is preferred by almost two thirds of the non sample users from regions outside Europe and outside the US, but only less than one third of the sample users. In addition, four fifths of the non sample users among interviewees from outside the US and outside Europe ruled out general consent, but only less than one fifth ruled out new informed consent.

No difference was found concerning those who used samples and those who did not concerning the acceptance of the policy in D.2. We did not find any patterns concerning respondents from the US sample who have been involved with biobank issues and the 11 respondents who had previously not worked on ethical questions related to genetic databases.

Otherwise it should be noted that some sample users were found in all groups, even in the groups (1) and (4). For example, in group (4) is one sample user working in Africa.

3. Discussion and conclusions

The reasons for agreement and disagreement in scenario D are similar to those in scenario A: different balancing between autonomy rights and costs, and benefits for research, different empirical assumptions about the burdens and benefits as well as about the feasibility of different consent types. Overall, in scenario D the disagreement becomes even more pronounced. The fact that general consent seems acceptable in a pure research context should not be extrapolated to samples taken in the context of medical care. In this scenario, the two fronts stiffened between those in favor of general consent and against new informed consent and those who hold the contrary views. The sizeable number of interviewees strongly believing that informed consent must be the prerequisite of any type of research amounts to a veto on a considerable number of research activities ("any individual whose material is being used by science has to consent. That's just a universal ethical norm," #63 US, bioethics, medicine, catholic theology, university, A). In particular, when it came to the question whether it is ethically acceptable to use existing collections for research without consent from the donors, but after approval by an ethics committee, a substantial number of interviewees answered "no." These findings are an indication of how much the problem of research use for hospital samples is influenced by the trust issue. Researchers themselves either trust each other to a great extent, or their own interest in not losing valuable samples makes them consider the risks of abuse less significant. Non researchers, especially outside the US and outside Europe and some of the non sample users in the US don't agree: they are concerned about "very sloppy research" (#14 IS, Oceania, political science, indigenous IP, university A/R) and convinced that it "happens all the time that the researchers can say that they have put a reasonable effort when they haven't done anything" (#77 IS, Africa, social sciences, NGO, A). Moreover, as already observed in scenario A, they don't trust local ethics committees sufficiently to let them decide about the use of hospital samples. The answers of the interviewees to scenario D illuminate this lack of trust in more detail. Some issues might have been raised in scenario D because they were brought to the attention of the interviewees during the discussions about the preceding scenarios. Examples are the idea of custodianship of the Ethics Review Board and the issue of collective consent which was mentioned more often in scenario D than in scenario A. For the issue of trust the following observations are of interest. In Europe and North America respondents seemed to identify the decision of an ethics committee with a form of collective permission, also referred to as custodianship of the samples, whereas respondents from developing countries and indigenous groups outside these regions made a clear difference between the decisions of ethics boards and the genuine instances of collective consent such as ancestor councils or other forms of tribal authorities.

Cultural differences seem also a major influence on the debate about consent and anonymization. In the US, where the fear of abuse is related to the risk of discrimination and stigmatization, anonymization is perceived as the appropriate means of eliminating these risks. Critics in this country arise from three directions: strong defendants of the respect for autonomy consider that the right of research participants extends to more than protection. It includes the right to know and to decide what happens to one's samples, including anonymization. The second group of critics consists of those referring to the limits of anonymity when it comes to DNA. Since it is not impossible, even though complicated, to identify a sample donor through DNA fingerprinting, consent is needed even for the use of anonymized samples. The third concern about anonymization is expressed by those who favor feedback of research results. In line with the principle of beneficence, they would want to give research participants the choice to receive or refuse feedback about individual research results that could influence their health or life planning.

In other regions where individuals see all bodily material even if detached from the person as part of the identity of individuals, their families and their tribes, anonymization is not a means to change the ethical issues related to consent.

3.2. Contributions to the current debate

Few international guidelines address explicitly the problem of research use of samples taken in a clinical context although the great majority of all samples worldwide are presently stored in pathology departments, most of them being part of hospitals and universities. The regulatory recommendations concerning this ethical and legal issue are usually buried in short passages on "research with samples stored for other purposes." However, our results show that arguments exist to support the claim that the original purpose of storage matters. Secondary uses of samples obtained for research do not raise exactly the same ethical issues as secondary uses of samples obtained during medical care. Interestingly, respondents from the international sample referred to a similar sliding scale as the HUGO guidelines (HUGO 1998): higher standards are proposed for the re-use of clinical samples than for research samples. However, while HUGO also required irreversible anonymization for the use of clinical samples without prior consent, this proposition was not accepted among many respondents from the international sample for the reasons explained above. In addition, although an opt out policy (i.e. a form of presumed consent as proposed by HUGO), was rejected less often in scenario D than in scenario A, such a policy is still opposed by many interviewees and ranks largely behind general consent.

International guidelines do not propose general consent openly, with the exception of recommendations from professional organizations such as the ESHG who suggests consent "for a broader use" (ESHG 2003, p. S9) and HUGO who considers "blanket consent" acceptable "in some cases" (HUGO 2003). The controversy about general consent in our study might therefore reflect the fact that respondents were aware of the guidelines. Those who used samples could have been influenced by the guidelines of their professional organizations whereas all others chose to give "politically correct" answers in favor of the traditional standard of informed consent. However, why would respondents from regions outside Europe and North America be more inclined to politically correct answers than respondents from these regions? One should rather take seriously the concerns of interviewees whose arguments refer to the realities in their country. The opposition to general consent and the opposition to the use of existing collections after reasonable efforts of recontacting have one aspect in common: the gatekeeper role of ethics committees who are supposed to allow only "ethical" research. The first step towards uniform international guidelines could be the implantation of trustworthy control mechanisms that fulfil local expectations of community representation. International organizations such as WHO could play a key role in such measures, which are in the end always dependent on available funds. The second step could be to work on uniform criteria that should govern the decisions of ethics committees. It is certainly no wonder that the "reasonable efforts" criterion is found predominantly in guidelines from Europe (COE 2006, ESHG 2003). One is tempted to conclude that while Europeans might be able to agree on a concept of "reasonableness," US Americans and other non Europeans are threatened by greater pluralism. Evidence is found in the ironical proposal of one US respondent that IRBs would need to show that no reasonable American will object to a particular secondary use of samples, which he adds would indeed be rare. The elaboration of criteria in international guidelines is important not only to increase trust, but also to obtain greater uniformity. As our study shows, respondents—many of whom are members of ethics committees or advise such committees—defended widely divergent criteria. Clearly, the consequences are large variations in the decisions taken by ethics committees, a fact also known from the literature (Silverman et al. 2001). Although it might be ethically acceptable that regional ethics committees use locally different criteria because they reflect distinct cultural values, such different standards make international research endeavours difficult if not impossible.

A surprising finding of our study is the opposition to research on anonymized samples without consent in the US sample, but also among some European respondents. Only a minority of interviewees support research on clinical samples in anonymized form (one third from North America and Europe and one fourth from all other regions). Major European guidelines (COE 2006) as well as regulation in the US permit this use and many research projects currently use this possibility. The existing guidelines might become more acceptable if it can be shown that after public notification, few research subjects would want to opt out of this type of research.

Our findings show significant opposition to the use of existing collections without consent. They are a reminder to researchers and hospital policy makers that they should urgently set up consent procedures for samples taken during medical care. At least some form of consent to research should be routinely obtained because the acceptability of later use is highly controversial.

Bibliography

COE (2006), *Recommendation Rec(2006)4 of the Committee of Ministers to member states on research on biological materials of human origin*, Strasbourg, France: Council of Europe.

HUGO [Human Genome Organisation] Ethics Committee (1998), *Statement on DNA Sampling: Control and Access* <http://www.hugo-international.org/Statement_on_DNA_Sampling.htm>, accessed July 2007.

HUGO [Human Genome Organisation] Ethics Committee (2003), "Statement on Human Genomic Databases. December 2002," *Eubios Journal of Asian and International Bioethics* 13, 99.

NBAC (1999), *Research involving Human Biological Materials: Ethical Issues and Policy Guidance, Report and Recommendations, vol. 1.* (Rockville: National Bioethics Advisory Commission).

SAMS (Swiss Academy of Medical Sciences) (2006), *Biobanques: Prélèvement, conservation et utilisation de matériel biologique humain* <http://www.samw.ch/>, accessed July 2007.

Silverman, H., Hull, S.C. and Sugarman, J. (2001), "Variability among institutional review boards' decisions within the context of a multicenter trial," *Crit Care Med* 29, 235-41.

UNESCO International Bioethics Committee (2003), *International Declaration on Human Genetic Data* (Paris, France: UNESCO).

World Health Organization (European Partnership on Patients' Rights and Citizens' Empowerment) (2003), *Genetic Databases, Assessing the Benefits and the Impact on Human Rights and Patient Rights*, Geneva.

World Medical Association (2002) *The World Medical Association Declaration on Ethical Considerations Regarding Health Databases* (Ferney-Voltaire, France: World Medical Association).

Chapter 7

Collective Consent

Agomoni Ganguli-Mitra

1. Introduction

In biomedical research, informed consent is considered a primary tool for the protection of research participants. Given that the potential for physical, social or psychological harm usually concerns each individual participant, informed consent has also traditionally concentrated on individuals. Yet for some time now, it has been recognized that the subject matter of genetic studies, as well as the isolated populations in which such studies are sometimes conducted, may make it appropriate to use something in place of, or in addition to, individual informed consent.

1.1 Special characteristics of genetic research relative to group consent

In all studies of a person's genes, the information produced has implications for that person's blood relatives and possibly for other members of that person's ethnic group. Thus in a way that is different than other types of biomedical studies, genetic research can make not just individuals but their families, communities and sometimes entire populations "subjects of research" (Greely 2001, 786). Moreover, research with the goal of learning about the genomes of distinct populations or the frequencies of particular genes of interest is research on the group itself rather than solely on those individuals whose genetic material is used. Sometimes, the process of conducting genetic testing may even serve to identify a cluster of people within a larger population who thereby become a distinct group. In all these cases, the risks and burdens related to genetic research may fall on the group rather than on individuals only, which raises the question whether it is necessary to find a means to get the group's permission for the research. This issue is even more pronounced when the genetic study involves a population that follows a traditional, collective form of decision-making rather than the Western norm of individual choice (Greely 2001, 787; Dodson and Williamson 1999, 204).

1.2 The current debate over group consent

Calls for taking into account the interests of groups in biomedical research first arose about 20 years ago, as it became more common for researchers from Western (industrialized) countries to conduct research in developing countries and in non-Western settings, where decision-making often revolves around collective rather than individual considerations of well-being (Weijer 1999, 501). In contexts where important decisions are taken as a community or under the advice of community leaders, researchers may find it advisable to consider the perspectives of the group as a whole. Some cultures, for example, may make little use of the notion of individual autonomy (McKendrick and Aratukutuku Bennett 2006, 67), at least in the sense commonly used in research ethics, and in such cases, relying solely on individual informed consent to make it licit to carry out research may be ethically questionable. Following the establishment of the Human Genome Diversity Project, which aimed to collect material from diverse populations around the world, various commentators suggested that existing ethical guidelines were too focused on the individual to provide adequate protection for individuals as part of groups (Weijer et al. 1999, 275; Kaye 2004, 121).

Yet, almost 20 years after the first calls to take collective interests into account, commentators remain divided on how to adequately protect group interests. The relevant literature contains various ways of describing collective involvement: collective consent, collective permission/approval, group/community/population consultation/involvement to name a few. It is often difficult to tease out an unequivocal normative approach as the terms themselves suggest both various levels of involvement and many ways of defining and designating groups. As to the former, "consent" appears to endorse more demanding measures and more explicit agreement than "consultation"; "involvement" can be interpreted as anything ranging from minimal consultation or mere dissemination of information to an ongoing, formal partnership at all levels of a study. Group, community, and population each suggest a different way of defining human associations that might require a collective approach. Even the notion of community, which might seem a relatively well-defined term at first glance, turns out to be "inherently fluid and porous" (Marshall and Berg 2006, 29). While Weijer et al., include in their concept of community "a wide variety of human associations including, ethnic, cultural, religious, political, artistic, professional and disease communities" (Weijer et al. 1999, 279). Davis would add to this list "stakeholders"; that is those for example, who are adoptive parents of a child from a certain ethnic community (Davis 2000, 40-1). Many commentators have pointed out that seeking collective consent can be difficult if not impossible precisely because it is very difficult not only to define the boundaries of a group but also to designate the appropriate representatives of groups (Greely 2001; Juengst 1998; Weijer et al. 1999). Juengst further points out that people create the boundaries of their groups according to many different characteristics and redefining those boundaries through genetic research can itself be detrimental (Juengst 1998, 192). Juengst, therefore suggests seeking group permission only where the social structure for collective consent exists. Others suggest making the requirement for group involvement less stringent, such as through procedures of consultation (Dickert and Sugarman 2006).

1.3 Various objectives for group consent

Behind these practical and terminological difficulties lie conceptual problems concerning the reasons for which group consent, permission, consultation, or involvement is sought. The objectives for consulting with, and seeking permission from, a group as a requisite for ethical research, range from the practical (that formal recognition of the group and deference to its leaders, such as village chiefs or tribal elders, is a prerequisite for having access to the group to recruit participants) to the ethical (that when the individual is not the unit of decision-making within a community, reliance on individual informed consent amounts to an imposition of alien moral premises on members of that community). Alternatively, group consent or permission is proposed when the objective is to protect the collective interests of the group which may be distinct from the interests of individual members of the group, and in any case cannot be waived by such members individually. In a related but distinct way, the interests may be individual but requiring group permission would serve to redress an imbalance in power between researchers and participants, just as workers organize to protect their individual interests through collective bargaining. Finally, requiring collective involvement or even formal permission may be seen as a screen, to protect prospective subjects who would, because of poverty or lack of education, not be able adequately to evaluate the risks and benefits involved in participating in a particular research project. Plainly, when people have different views about the goals that could be served by taking a group rather than solely an individual approach to consent for research, they are likely to disagree about what role the group should play and how the group should be defined and identified.

As noted by Vayena et al. in chapter 3, research guidelines follow the traditional ethical principles of individual consent, even if in certain cases, some form of group involvement may be desirable, for example in cases where research carries a risk of stigmatization for the entire group. It is however, interesting to see that despite the nature of genetic research, few of the international guidelines on genetic databases or biobanks address this matter in detail, with exceptionally, the HUGO (2002) guidelines mentioning the role of communities in research (see chapter 3.2.2). This topic, however has been taken up by some regional and professional bodies such as the European Society for Human Genetics, ESHG and the Canadian Science and Technology Ethics Committee (CEST), both in 2003 in their recommendations on DNA banking (see chapter 3.2.2; Boggio et al., 2005, 5-6). The CEST, which addresses population-based genetic databases, recommends a public consultation in five steps, starting from a population survey and culminating in a policy statement that takes into account all stakeholders and interest groups (CEST 2003, Rec. 15). Beyond such specific guidelines, however, the issue of collective or group involvement has not been adequately addressed by international frameworks.

1.4 How the topic was addressed in this study

In our project, the topic of consent cut across several of the scenarios, especially Scenario A, while the topic of group involvement was addressed in a relatively contained manner in Scenario C. In this scenario, researchers wishing to collect samples for commercial pharmacogenetic studies approach an indigenous population; the size and location of this population are unspecified, other than being located in the same country as the researchers, thereby eliminating consideration of a transnational search for genetic material. Having identified the group's representatives in the form of a governing council, the researchers first approach them as gatekeepers to the community. Negotiations regarding benefit-sharing fail (see chapter 12) at which point the governing council refuses, on behalf of the entire population, to participate in research. The researchers then consider whether to bypass the governing council and approach individual members for participation in exchange for US $800 as remuneration. Interviewees were asked whether such an approach was adequate, and which characteristics, if any, required collective consent.

2. Results

2.1 Practical reasons for collective consent

With certain exceptions and nuances, the results of the interviews largely reflect the ongoing debate in the literature. Certain respondents, although few in number, approached the collective issue as a practical matter rather than an ethical issue. Where traditional structures of collective consent or a system of group representatives are in place, it is practically less troublesome for researchers to follow them; indeed, in some circumstances that may be they only way for the researchers to obtain access to the setting where they want to carry out their research. Some respondents felt that obtaining the governing council's permission was more judicious or advisable (#24 IS, Europe, medicine, consultancy, A/R) either because the researchers might otherwise destroy relations with the group or because research would otherwise be impossible. This was true even for certain cases where it was felt that in principle, such a thing as "group rights" should not be recognized. Similarly, some respondents felt that going against the wishes of group representatives could be disruptive and possibly create antagonism among group members (#28 US, bioethics, medicine, government, NI; #54 US, bioethics, protestant theology, philosophy, university, R/A) and therefore would be an imprudent decision. It is not clear whether such a reaction was prompted by the particular situation in this scenario, where researchers considered going to individuals despite the failed negotiations with the governing council, a gesture some found dishonest or "sneaky" (#41 US, bioethics, law, university hospital, R/A). It is noteworthy however, that these respondents encouraged collective consent for very different reasons than those respondents who came from a cultural background where such practices are common and who felt for example that a collective decision in this context would be analogous to a decision concerning communal property, such as land (#77 IS, Oceania, social sciences, NGO, A; #37 IS, Oceania, social sciences, bioethics, university, ERC, A).

2.2 Principle-based reasons for collective consent

Among the many respondents who endorsed some form of collective decision-making, few referred solely to practical reasons but rather many also advanced various arguments based on principle. In sharp distinction to those who reject group rights were those who believe that such a procedure was necessary in order to respect group identity or what one respondent called "group autonomy" (#65 IS, So. Am., life sciences, genetics, university, A). Beyond the principle of collective identity however, traditional structures of collective consent are also, some proposed, an important mechanism of individual protection:

> People are vulnerable. These are communities that are community-oriented. Individuals are very vulnerable in themselves, unlike the West where historically the individual has been protected by constitutions, awareness, and culture (#61 IS, Africa and Europe/Africa, life sciences, university, R/U/A/O).

When asked which characteristics required collective consent, many respondents suggested that such a protective mechanism should be respected whenever the population studied was vulnerable and therefore prone to exploitation. Such vulnerability could be manifested through various characteristics, including poverty, "You get people to do everything for a dollar and that does not seem fair" (#25 US, philosophy, government, N), and lack of education, "It depends on which societies. If people are educated, then you have individual consent. But if the people are not educated, they don't understand what you are talking about" (#78 IS, Asia, life sciences, government, S/A). Others felt that even a certain level of education provided little awareness of the consequences and risks of genetic studies. Thus, a collective approach was needed because it would act as a safeguard for individual interests:

> In some cases it is necessary, especially in indigenous community, with information that is very scientific. It is a question of whether an individual person is clearly aware of all the implication of the investigations (#79 IS, So. Am., medicine, university, S/A).

2.3 Processes for collective consent

While acknowledging the importance of traditional structures, many respondents took a weaker stance on collective consent. They emphasized rather the importance of taking cultural factors into account in carrying out research (#36 US, medicine, genetics, government, R/A/S), for example, by engaging in public or community-wide consultation processes:

> It depends on the values of the society. We don't want to impose values that are foreign to a society; we don't want to break up a society (#11 US, medicine, bioethics, university, R/A).

A public consultation or deliberation, according to some, would provide the appropriate cultural approach and set the scene for a subsequent, adequate individual consent:

> It is appropriate to have collective permission, some kind of political permission for instance; it is important to have public deliberation but it doesn't replace and can never replace individual consent. It is more in the background. The political debate provides the background when individual make up their minds (#48 IS, No. Am./Europe, philosophy, university, R/A).

Other respondents who also favored group permission did so on very different premises. These respondents feel that in cases such as scenario C, individual consent is in fact of much lesser importance:

> What we do is that in our communities ... in the most remote areas, we agree that group consent should be taken. The leaders must decide for the group. If they agree, then their consent should be taken. The indigenous consent is less useful. The individuals are not giving consent. The group permission should be taken (#74 IS, Asia, genetics, hospital, S).

The approach of public deliberation or consultation is also popular in the existing literature (Dickert and Sugarman 2006, 26). Some of the interview responses also echoed the calls in the recent literature to shift the emphasis of consent from the formality of documentation to correct process. However, it is not clear whether such a consultation should take place even when the negotiation with representatives has broken down, or merely as a supplement to it. The disagreement still seems to persist regarding whether the governing council in this case has the right to veto individual consent as it does with other decisions within the community. This problem is amplified when an established structure of collective decision-making does not exist, so that the rationale for seeking collective consent is instead the shared nature of genetic information. A few respondents felt that this shared nature was reason enough to seek collective consent. One view among those respondents who favored collective consent because of the nature of genetic information, was that the need for collective consent arises from the nature of the study itself:

> When researchers decide they want to research indigenous groups, the consent starts at that stage, right at the outset, in the design of that program. That's how you begin to build a constructive relationship; you start at the very outset. If you wanted to research individuals, it is slightly different. If you say that you want to go and research the Navajo or Polynesians or people in Tonga, what you are saying is that you want them because they are part of the collective. Then you must find their collective body, and if there isn't a collective body, you have to make a choice to either go and collect samples somewhere else or to work with national institutions to try and find a way to get a collective [permission] (#14 IS, Oceania, political science, indigenous IP, university A/R).

Another group of respondents took a slightly different approach and agreed that such a requirement should be in place only when a group of participants in research might potentially be affected by the results of that research (#46 IS, No. Am., law, social science, NGO, A). A few of these respondents suggested that the requirement of collective consent could be limited just to groups that can be clearly labeled and that potentially face discrimination and stigmatization as a result of the research (#12 US, Africa, genetics, university, R/A/S).

Interestingly, only a few of the respondents who endorsed collective consent for genetic research also consistently expressed such positions when responding to other questions in this interview, though some did. For example, when asked about the fate of samples upon a participant's death under scenario A (which raises ethical issues about disease-specific databases), one respondent suggested that the family might be given the right to control the use of samples just as they would other aspects of "family heritage," while another (quoted above as supporting collective consent due to the nature of genetic studies) said that such decision should be taken by a donor association:

> Here again the issue of collective consent comes in. In my view, the collective group then becomes the protective mechanisms for the donor, as in some situations where you can have a Donor Association, so that they are kept in touch with each other and updated on the progresses of the research (#14 IS, Oceania, political science, indigenous IP, university A/R).

It is clear that while scenario C directly addressed the idea of the collective, others that also involve shared genetic information, (such as scenario A) may not have evoked it directly. It is possible that those respondents whose endorsement of a collective approach rests on the shared nature of genetic information also rely on some kind of structure: a structure delineated by research design, a social (family) structure or an organized structure such as a patient group or donor association.

2.4 Characteristics favoring group consent

As predicted by the debate in the literature, there was little consensus regarding which characteristics in a population systematically required collective consent. Applying standardized norms to all groups may, arguably, be difficult and undesirable, given that certain combinations of characteristics would require collective consent while others would not. The questions regarding group characteristics were the ones on which respondents were the most divided. The concerns raised here were very similar both for respondents who favored collective consent and those who advised against it. While the latter used these concerns as their main arguments against collective consent, the former tended to treat them principally as obstacles in seeking it. One of the most frequently stated practical concerns was the problem in identifying a community or group and of defining its boundaries. This is especially true where the study design dictates the need for collective consent. While it is relatively easier to define the boundaries of a tribe, this is not the case for larger populations such as the Ashkenazi Jewish population. The exercise seems even more difficult when groups are defined with characteristics other than ethnic origins or geographical boundaries:

> I would be inclined to go by characteristics, collective permission for groups that share the characteristics that you would like to study ... Not all ... you can't just talk to all the people with heart conditions if you want to study heart conditions (#25 US, philosophy, government, N).

Population size and level of heterogeneity and cohesion, among other things, seem to pose a rather important obstacle for collective consent:

> The individual approach is important. Even if it seems a coherent group, they are heterogeneous. It is a hard question also in [my] society ... I think [the individual approach is dominant] because if you say 'groups' it is very tough to imagine actually (# 67 IS, Asia, genetics, government, biobank, R/A).

2.6 The adequacy of group consent in representing individuals

Even where it is possible to define a group or community adequately, we cannot be sure that collective consent or permission from community representatives would protect all individuals affected by the study. It would not include those who have left the community, or who choose not to be represented by their community leaders. Yet the study results would affect them equally. As pointed out by one respondent, it is as important to recognize group hostilities as it is group affiliations (#59 US, medicine, bioethics, government, R/A/S). Closely related to this issue is the task, as a first step, of identifying the representatives of a community and, as a second step, assessing whether they adequately represent their community. The potential for misrepresentation might be one of the most delicate ethical issues in this context. As it was often pointed out during the interviews, by seeking consent from representatives, instead of protecting the interests of each individual member of the community, one might unknowingly reinforce existing power imbalances and the exploitation of already weaker members within the group:

> I would want to know lots more about this group and this council. There are governing councils in developing countries who have misrepresented, abused and exploited their own people for a long time so that could be what's going on. On the other hand this might be appropriate, legitimate representatives of the population. So if it is the second thing, they can't do this. If it is the first thing, maybe I [would] consider it (#27 US, philosophy, bioethics, government, R/A).

To partly remedy this problem, one respondent suggests relying on a third party, such as an Institutional Review Board (IRB, or research ethics committee), to act as an additional gatekeeper, though it is unclear whether an IRB would truly be able to pinpoint the symptoms of misrepresentation if the IRB is mainly composed of members foreign to the culture. On the other hand, involving a mediator from the same culture as the potential research participants and their representatives, may potentially present similar problems as with the governing council. As suggested by a few respondents, one may have to "take a step back" and undertake some cultural assessment before being able to develop a culturally and ethically adequate approach (#26 US, medicine, social sciences, government, R/A).

2.7 Notable similarities and differences between subgroups of respondents

The various issues and concerns regarding collective consent raised by respondents were not specifically associated with the respondents being from one particular background or another. Nonetheless, respondents who came from cultures where collective decision-making is a widespread norm more readily favored such a procedure; however, many of them also warned against the possibility of representatives exploiting their own people. Respondents from the US as well as Western respondents in the international sample, were often more reluctant to address the issue of collective consent "head on." Some preferred to skip the question because they felt that they lacked the adequate expertise to answer it. Sometimes responses suggested that the interviewee viewed the situation as an "outsider" who was not entirely at ease with the cultural context of the scenario. Other respondents in this group while addressing the issue were less willing to accept collective consent unconditionally. In general they tended to prefer a more open approach, including a cultural assessment and an assessment of the legitimacy of representatives, as well as processes of consultation and public deliberation.

3. Discussion and conclusions

Although the results of the interviews suggest that no consensus exists on all points on the topic of collective consent, it is important to note that very few of the respondents favored ignoring the group perspective altogether and relying strictly on individual consent. What still remains nebulous are which characteristics of a particular situation are seen as making collective input practically or ethically necessary as well as how extensive such collective involvement should be. While some respondents feel that only an existing traditional structure of collective decision-making warrants making genetic researchers seek collective consent, many would also favor collective consent when there is the potential for discrimination arising from the results of the research or when participants lack education or economic power. Respondents also echoed the concerns expressed in the literature regarding the procedural aspects of collective involvement such as designating and defining the boundaries of a community, recognizing the legitimate representatives of a group and upholding the rights of all those affected by the research.

An interesting question to ask is whether collective consent and individual consent can truly coexist in a given population. In settings where individual autonomy is of primary importance, or where many members of a group are located far away from each other, it may be possible to have a widespread public consultation, followed by individual informed consent. However, in settings where researchers have sought and obtained the permission or consent of leaders, it is worth asking whether individual consent does not then become a futile exercise. The more traditional the setting, the easier it is for researchers to define the representatives, but it is also more likely that in those settings, community members will not be willing to depart from a decision made by the group leaders (even if the leaders' decision is simply that the research may proceed rather than formally granting "consent" for each individual). Conversely, once group representatives have vetoed participation in research, it may be in the interest of all that no one participates. This may be particularly true when, as in the case of genetic research, the reason for seeking consent from the group is to recognize that, in addition to individual risks and burdens, the results of the research may have direct implications for the collective interests of the group itself. Of course, it then becomes difficult to rely on the legitimacy of the governing council or self-proclaimed group leaders.

The results also show that in order to work towards a global consensus on collective consent for biobanks, further in-depth discussion is required on both the theoretical and procedural aspects of collective involvement. Beyond educating the researchers or the review board on the important cultural aspects of various research settings, which may help them in choosing legitimate representatives where collective involvement is required and in detecting the symptoms of misrepresentation, few other measures seem currently universally applicable. Instead, investigators and sponsors carrying out genetic research are either left with taking the partnership approach, leaving it up to the community to decide for itself while risking reinforcing existing oppressive customs, or adopting a paternalistic approach and refusing to carry out research unless they feel absolutely sure that basic rights are not violated. It may well be that a uniform regulation of collective involvement is not desirable, that both the reason for seeking collective involvement and the exact procedural approaches must be flexible according to various contexts. It is, however, necessary to work towards developing alternative solutions based on existing needs, solutions that stem from a common drive to carry out research that protects and benefits all its stakeholders.

Bibliography

Boggio, A. Biller-Andorno, N., Elger, B., Mauron, A. and Capron, A.M. (2005), "Comparing guidelines on biobanks: emerging consensus and unresolved controversies" <http://www.ruig-gian.org/research/outputs/output.php?ID=254>, accessed 17 September 2007.

CEST, Commission de l'éthique de la science et de la technologie (2003), "Les enjeux éthiques des banques d'information génétique: pour un encadrement démocratique et responsable" http://www.ethique.gouv.qc.ca/Les-enjeux-ethiques-des-banques-d.html#documents. Accessed on 14 August 2007.

Davis, D.S. (2000), "Groups, communities and contested identities in genetic research," *Hastings Centre Report* 30: 6, 38-45.

Dickert, N. and Sugarman, J. (2006), "Ethical Goals of Community Consultation in Research," *Am J Public Health* 85: 1123-1127.

Dodson, M. and Williamson, R. (1999), "Indigenous peoples and the morality of the human genome diversity project," *JME* 25, 204-8.

Greely, H.T. (2001), "Informed consent and other ethical issues in human population genetics," *Annu. Rev. Genet.* 35, 785-800.

Juengst, E.T. (1998), "Groups as gatekeepers to genomic research: conceptually confusing, morally hazardous, and practically useless," in *Kenn Ins Ethics J* 8(2):183-2000.

Kaye, J. (2004), "Abandoning informed consent: the case of genetic research in population collections," in Richard Tutton and Oonagh Corrigan (eds), *Genetic Databases: Socio-ethical Issues in the Collection and Use of DNA* (London/New York: Routledge), pp. 117-138.

Marshall, P.A. and Berg, J.W. (2006), "Protecting Communities in Biomedical Research," *AJOB* 6:3, 28-30.

McKendrick, J. and Aratukutuku Bennett, P. (2006), "Health research across cultures-an ethical dilemma?," *Monash Bioethics Review* 25:1, 64-71.

Weijer, C. (1999), "Protecting communities in research: philosophical and pragmatic challenges," *Cambridge Quarterly of Health Care Ethics* 8: 501-13.

Weijer, C., Goldsand G. and Ezekiel, E.J. (1999), "Protecting communities in research: current guidelines and limits of extrapolations," *Nat. Gen.* 23, 275-80.

Chapter 8

Withdrawal of Consent and Destruction of Samples

Bernice Elger

1. Introduction

Withdrawal of consent and the destruction of samples and information is another issue stirring controversy in the context of biobanks. The attitudes of the respondents from our study towards this issue are particularly interesting since the literature on these topics is less abundant than on other details concerning biobanks.

1.1 Withdrawal of consent and destruction of samples and information: the debate

The right of research subjects to withdraw from a study was stipulated in the Nuremberg Code and has since then been a very important part of classical research ethics.

> During the course of the experiment the human subject should be at liberty to bring the experiment to an end if he has reached the physical or mental state where continuation of the experiment seems to him to be impossible (Nuremberg Code 1947, art 9).

What does withdrawal mean? The original understanding was that research subjects are allowed to "walk away." One could not keep them against their will in the hospital or force them to take experimental drugs. The question of whether this right of withdrawal is applicable to the material or data already obtained from research subjects—that is, whether subjects may withdraw not only themselves but also biological material and personal data that they have already provided to researchers—has been raised only recently, under the influence of the debate about biobanks. Researchers are worried because they fear that research participants exercising a right to withdraw by having their data and samples destroyed could bias research results. For example, patients who withdraw are often those who have experienced adverse effects, and their data are crucial to show negative effects of the medication being tested.

Since the right to withdraw was originally formulated as a means of ensuring that research subjects remain free to limit their participation based on their immediate suffering in an experiment, some argue that the right does not exist for research involving samples or information because the donors are not directly burdened by researchers' use of their data, detached body parts or other biological material. Several counter-arguments have been raised, however. First, the concept of withdrawal based on gaining release from suffering is too narrow even as regards research involving subjects directly. For example, a new medication being tested in pharmaceutical research could be dangerous although research subjects don't suffer any adverse effects immediately. Reasonable persons would want to withdraw even if adverse effects such as pulmonary fibrosis or liver problems could develop only in the future. Second, some adverse effects might be related to the subjects' wish to avoid violations of privacy or other non-physical harms that could arise from their material or data being used in research. To be able to withdraw, research participants have to receive adequate information about factors that might influence their decision to remain in the study, such as additionally discovered adverse effects or risks which had not been specified in the consent form. Finally, if the right to withdraw is simply a way of saying that consent to research is a continuous process rather than a one-time event, then subjects remain free for any reason simply to declare that they no longer wish to participate in a research project, and in the case of biobanks their "participation" consists of further use of their data and biological materials that are held by the biobanks. Respect for autonomy implies that part of the right to withdraw from research is the rule that research participants don't need to give a reason for their choice not to participate further in a study. Any other policy would represent forced, and therefore unethical, participation.

The leading statements on research ethics exhibit the ambiguity regarding the application of the right to withdraw in certain situations. For example, the Declaration of Helsinki could be interpreted in at least two ways. Article 22 states that the "subject should be informed of the right to abstain from participation in the study or to withdraw consent to participate at any time without reprisal"(WMA 2000). This language could be interpreted as recognizing a right to stop participation as of the moment of withdrawal. In the case of research on samples or information, that would mean that all data and samples gathered until this point would remain with the researchers. However, the term "participation" could also be interpreted more broadly as meaning that if research is carried out on their samples and information, subjects are still "participating" in research because their data will be analyzed in the future and they remain exposed to risks (e.g., privacy violations). If withdrawal is intended to protect against harms of this sort, withdrawal of consent could be effectuated through the irreversible anonymization of samples and data. However, others argue that the principle of respect for autonomy implies that research participants should be able to mandate the destruction of their DNA samples (Harlan 2004, pp. 179-219) and of all unpublished data generated from them, since anything less than this means that participants are still (involuntarily) involved in the research.

Rules about destruction of samples and data obviously affect biobanks. The issue of destruction is not only relevant in the context of withdrawal, but also at the end of a research project or when funding of the database or repository runs out. In the following part, we will summarize international recommendations on withdrawal and destruction of samples and data in the context of genetic databases.

1.2 Withdrawal of consent and destruction of samples and information: published recommendations

International guidelines agree that a right of withdrawal should be granted, but do not agree on all details. Consensus exists on the condition that "withdrawal of consent should entail neither a disadvantage nor a penalty for the person concerned" (UNESCO 2003, Article 9). Similarly, the Council of Europe stipulates that the "withdrawal or alteration of consent[1] should not lead to any form of discrimination against the person concerned, in particular regarding the right to medical care" (COE 2006, Art. 15).

Guidelines disagree concerning the right to order destruction of samples and information after the withdrawal of consent. UNESCO foresees the possibility for the researcher to use samples and data after withdrawal as long as they are "irretrievably unlinked to the person concerned" (UNESCO 2003, Art. 9). In line with UNESCO, the Council of Europe does also not seem to grant a right to destruction but only a right to irreversible anonymization:

> When identifiable biological materials are stored for research purposes only, the person who has withdrawn consent should have the right to have, in the manner foreseen by national law, the materials either destroyed or rendered unlinked and anonymized (COE 2006, Art. 15).

By contrast, the WMA's declaration on health databases stipulates the right to destruction of data contained in a database:

> Patients should have the right to decide that their personal health information in a database (as defined in 7.2) be deleted (WMA 2002, Art. 10).[2]

Similarly, the European Society of Human Genetics mentions that research participants have the "right to withdraw at any time from the research, including destruction of their sample" (ESHG 2003, p. S9). The WHO guidelines, although generally in favor of a right to withdrawal including destruction, permit exceptions to this right if "the holder of the information or sample [is able] to show that it is not reasonably practicable to comply with this request" (WHO 2003, 6.2).

Should research participants be allowed to waive their right of withdrawal? This issue is discussed only in the WHO guidelines, which stipulate that the right to withdrawal of samples and information "is not waivable by consent, except where absolute anonymity is guaranteed" (WHO 2003, 6.2).

The most explicit restrictions on the right to withdraw appear in the HUGO guidelines. Although it begins by recognizing that the "possibility of withdrawal of consent" without prejudice is an ethical prerequisite for medical research, HUGO states that the individual wishes of sample donors with respect to destruction of

---

1 The Council of Europe contains not only a right to withdraw consent, but also a right to change the scope of the original consent (COE 2006, Art. 15).

2 Art 7.2 provides the explanations that the definition of "database" "does not include information in the clinical record of any individual patient" (WMA 2002, Art. 7.2).

their samples may "occasionally" be outweighed by "special interests and moral obligations' of biological relatives of participants who share genetic risks (HUGO 1998). Therefore, "stored samples may be destroyed at the specific request of the person" only in "the absence of need for access by immediate relatives" (HUGO 1998). Moreover, HUGO adds another exception to the right to withdraw based on the negative consequences for research projects or clinical diagnosis. Destruction is "not possible for samples already provided to other researchers or if already entered into a research protocol or used for diagnostic purposes" (HUGO 1998).

None of these international guidelines explicitly discusses the destruction of samples at the end of a study. Only the ESHG document contains a requirement to inform research participants about the "duration of storage" of their samples and information (ESHG 2003 p. S8). UNESCO mentions this issue only with respect to data and biological samples collected for forensic purposes. Data and samples obtained from a suspect in the course of a criminal investigation "should be destroyed when they are no longer necessary, unless otherwise provided for by domestic law consistent with the international law of human rights." Similarly, genetic data and samples "should be available for forensic purposes and civil proceedings only for as long as they are necessary for those proceedings, unless otherwise provided for by domestic law consistent with the international law of human rights" (UNESCO Art. 21).

The Council of Europe mentions this topic only in its articles concerning population biobanks and does not require anything more than that "[p]rocedures should be developed for the transfer and for the closure of a population biobank" (COE 2006, Art 19.3).

The guidelines imply that in many cases indefinite storage should not be prohibited for the public good: "Repositories should be established and funded to ensure the continuation of publicly available databases" (HUGO 2002, Art. 3(b)). Furthermore, NBAC offers the reminder that the issue of destruction of samples may be influenced by legal requirements. Patients may not realize that federal and state regulations require that specimens be stored for a certain length of time because protocols and procedures that are followed in collecting and disseminating samples might not have addressed issues regarding destruction of the materials (NBAC 1999, p. 7; Merz 1996, p. 7f).

Guidelines are silent concerning the use of samples and information and family members' possible right to withdraw samples and data after the death of a research participant. They seem to suggest that this issue is dependent upon what has been specified in the consent form or is regulated according to the legal context of various countries. In most Western European regulations, the wishes of patients concerning their bodies outweigh the wishes of relatives. In legal terms, consent to research use, if not specified otherwise, might automatically authorize the use of samples after death.

1.3 Which policies are adequate concerning withdrawal of consent and destruction of samples (scenarios A.2 and A.8)?

The issue of withdrawal and destruction of samples and information was addressed in scenario A.2. Respondents were asked to indicate which of four withdrawal policies would be the best and which policies they would rule out. Option (a) limits the right of withdrawal to research participants abstaining from providing any information and samples in the future. No right to withdraw exists for samples and associated information already submitted to the repository. Option (b) restricts the right to withdraw to requesting irreversible anonymization of samples and information. Research would be allowed to go on using the anonymized samples and information. Option (c) states that withdrawal implies the destruction of any samples and associated information contained in the biobank. Finally, option (d) extends the destruction of samples and data further in order to include not only the material stored in the biobank, but also any samples and information provided to researchers. The latter "will be obliged to destroy the samples and to remove the samples and associated information from any report of research that has not yet been submitted for publication."

In scenario A.8, two alternative policies are described regarding destruction of samples at the end of a study. One involves closure of the repository and destruction of samples and associated data, "once the project has completed its study on colorectal cancer." The proposed alternative is to maintain the repository and not to destroy the samples and associated data even after the colorectal cancer study has ended, as long as the country where the repository is situated is willing to maintain it.

2. Results

2.1. Arguments in favor of and against different policies on withdrawal

Concerning the options regarding withdrawal, the answers of respondents from Europe, North America and all other regions were similar and no difference was found between the attitudes of respondents who have used or stored samples and those who have not been involved directly with samples. The policy most often chosen is destruction of all material including what has been sent to researchers (option d). This option is favored by more than one third of respondents, but also ruled out by another third of interviewees. Destruction of all information in the biobank (option c) follows closely. It is preferred by one fourth of respondents and is the policy they ruled out least often (less than 20% of interviewees). One fifth of respondents favor anonymization of samples in the case of withdrawal (option b), but almost half of respondents ruled this policy out. Less than 15% favor no right to withdraw samples and information that have been already provided to the repository (option a). This option was the policy most often rejected (more than half of all respondents). Overall, the more restrictive the withdrawal policy, the less often it was preferred by the interviewees included in our study. Only a few respondents from North America opposed all options offered in scenario A.2. They think that withdrawal should not be allowed at all.

Information provided to research subjects and different forms of the right to choose Several respondents underlined that withdrawal and its implementation must be presented to research participants in a clear way. "The important issue is how withdrawal is presented to people" (#53 IS, Europe, medicine, public health, Int. Org., R/A/S). The consent form should alert research participants in particular if withdrawal is not permitted or if withdrawal would entail irreversible anonymization of samples and data and their continued use.

> What it is important is to be clear. 'You don't have a right of withdrawal,' then the participant knows what the offer is and he can accept it or not, although people would be more comfortable to have a right to withdrawal if something catastrophic seems to happen—the data are used for purposes that don't seem to be right or have been sold to commercial companies. I don't have inherently any objection to any of these options as long as people know what they are guaranteed (#68 IS, No. Am./Europe [nationality: No.Am.], life sciences, consultancy R/A^3, in favor of option d).

Several interviewees found it difficult to decide between the options because "there are plusses and minuses to all of them" (#96 US, genetics, medicine, social sciences, university hospital, R/A/S). A sizeable number said that all forms of withdrawal or at least several of them are acceptable as long as the type of withdrawal is explained to research participants.

> [Options] (b), (c) and (d) all seem reasonable and so they are gradations of how much you care about it (#16 US, philosophy, bioethics, university, N).

> This is always subject to the consent form and that the subjects really understood when they agree to participate in the study. I think that's the most important principle from my perspective that there is clear consent and an understanding of how the sample would be used (#47 US, law, social science, government, A).

Some participants expressed the view that all options must be available to participants so that they can choose the one they prefer because "this decision should lie with the individual donor" (#46 IS, No. Am., law, social science, NGO, A).

> Ideally the participant is the best [person] to answer this question for everyone. I cannot really answer for the participant. You know if you involve a family in the study, I don't know how they will behave let's say after one year (#74 IS, Asia, genetics, hospital, S).

In fact, as one respondent suggested, "patients can make quite important decisions themselves if they are given a range of options ... people's relationships within the scientific research change. So there is need to be flexible" (#14 IS, Oceania, political science, indigenous IP, university A/R).

Arguments against all forms of withdrawal   One respondent thinks that research participants should have more options than being able to choose between only four types of withdrawal. This interviewee is against each single withdrawal policy

---

3 The respondents' different degrees of involvement with biobank issues are noted: R= making recommendations, S = sample use and/or storage, A = analyzing ethical and legal issues, O = other involvement, N = respondents with no direct involvement with biobank issues.

because he thinks that research participants should be able to choose their own personal conditions concerning withdrawal.

> There needs to be a flexible contract with rights and responsibilities of everyone involved. [Even policy] (d) does not address the issue of equal partnership. It is just a passive way of saying no (#51 US, Europe, bioethics, philosophy, theology, O).

Another respondent argued that autonomy is sufficiently respected if participants are informed that no right to withdrawal will be accorded at all. Research participants have the opportunity to decide whether they would want to withdraw and if this is the case they should simply not provide any sample at all to the biobank. This interviewee is convinced that withdrawal would hamper good research.

> I think there should not be any withdrawal allowed because it is such a fundamental study that this would confuse and create problems. I think the individuals should be informed at the very beginning that once they give their consent this is final and that there is no allowance for withdrawal. And then if they know that and think for some reason they might want to withdraw then they don't give their consent, they don't give their sample and they don't enter into the repository. I don't think you can have good research with people withdrawing samples at whim (#63 US, bioethics, medicine, catholic theology, university, A).

This interviewee's disagreement with any policy of withdrawal, including policy (a), goes beyond the data already held by a biobanks. In his opinion, once research subjects have consented to the study, they should not be allowed at any time in the future to stop the information flow from their physician to the biobank.

> [I]f the individual knows at the beginning the conditions under which he gives consent …and says 'yes' then it is ok…When he hears…that he will be contacted from time to time and his physician will be contacted to get information and he decides 'oh I am not sure I want this' then he should not be in the project. He could refuse to want to hear anything back from the project but he can't refuse his physician from pushing individual information into the project (#63 US, bioethics, medicine, catholic theology, university, A).

Another respondent, also from the US sample, thinks that no right of withdrawal could be justified provided that the conditions guarantee the protection of research subjects.

> If full disclosure is made to the patient that he cannot withdraw this might be acceptable. The repository should then have an obligation to have a trustee to protect patients' rights and their interests to know results (#11 US, medicine, bioethics, university, R/A).

Arguments in favor of option (a): no right to withdrawal of samples and information provided in the past   Respondents who favor option (a) said this withdrawal policy is superior to all others because samples would not be wasted and adverse effects for research can be avoided.

I would say I am for (a). (c) and (d) I find is destroying what has been done. That will not be good for the research, a waste (#05 US, Asia, philosophy, bioethics, theology university, N).

A concept used frequently to defend both policy (a) and policy (b) was "fairness" to research participants (#47 US, law, social science, government, A). Policy (a) is seen as the best way to balance practicability with fairness to participants.

I think that (a) might be a reasonable compromise ... I think that a fair and practical solution is that once they have given the samples the material that they have given might be used but not for new projects (#28 US, bioethics, medicine, government, N).

Fairness seems to include two elements: that research participants have been informed in the consent form and that they are not exposed to any significant risks by the policy.

If (a) is in the consent form [it is ok]. I am trying to think what the risk of harm is to the person: I would not have a problem with (a) (#01 US, medicine, bioethics, university, N).

Respondents in favor of option (a) think that consent is sufficient to waive the right of withdrawal for samples and information once they are stored in the biobank.

When it is not acceptable to them and it is clearly part of the consent process, then they would choose not to participate, and that's fine. So in some ways I feel as long as the consent process makes very clear what the researcher's intentions are and people understand that they have the right to participate or not at that point and they agree to participate, then I am not so worried about [taking away from] them the right to withdraw (#07 US, social sciences, bioethics, university, O).

I think that policy (a) is the preferable one. I am not in favor of creating a right to withdraw. I am in favor of consenting people in advance so that they understand it is impractical to withdraw (#98 US, philosophy, medicine, genetics, university, R/A/S).

As shown by several of the previous answers, many respondents who chose policy (a) are convinced that withdrawal is difficult or even impractical.

Several respondents, in both the international and US samples, used economical or legal terminology to express why consent should be binding for research participants: they agreed to a contract that they are obligated to fulfil. Participants ought not to be allowed to breach the contract if this causes harm to the research.

[I favor option (a)] because when you give your consent and your sample, it is disruptive to the study if you later change you mind and want to withdraw the sample. So I think, a deal is a deal (#54 US, bioethics, protestant theology, philosophy, university, R/A).

I would vote for (a) if consent has been given. If uses are disclosed and there is informed consent, I don't support actually participants having the right to withdraw samples. ... Once there is consent, if there is authorization, the commitment is a contract ... If that is the case, I don't see why participants would withdraw consent. It is extremely arbitrary, and I don't think that is what I would like to support (#21 IS, Europe/Asia, philosophy, university, A).

In addition, interviewees believe not only that withdrawal is impractical, harmful or burdensome for the biobank, but also that reasonable participants would not have any reasons to withdraw since no significant harm is associated with biobank research (#01 US, medicine, bioethics, university, N).

One respondent acknowledged that policy (a) might discourage research participants to consent to the study. However, this interviewee shared the view expressed by others that withdrawal represents an important threat to the correct functioning of biobanks. Therefore, in his view, it is better for the biobank to contain fewer samples, provided that all come from participants who have consented to policy (a):

> I would rather have a less comprehensive bank than trying to track and deal with withdrawals. I don't think banks can be operated this way and deal with that [withdrawal of samples], people move and die and marry or change names (#98 US, philosophy, medicine, genetics, university, R/A/S).

Finally, one respondent explained that policy (a) complies best with the right to withdraw as it has been conceived in traditional research ethics. According to this respondent, it is the right to abstain from all future interventions on the person, but not the right to destroy anything that has been provided to researchers in the past. Hence he thinks that policy (a) corresponds to the "usual" meaning of withdrawal. He assumes that in line with this usual understanding, research participants would not want to destroy their samples and information, but just to "stop coming."

> [I am in favor of policy (a)] because I think that, first, it is the easiest to operate; second, it adequately respects in general the wishes of subjects who choose to withdraw from research ... Most people, when they want to withdraw from research, they want to stop coming. There is no reason to think that a person would ever want to destroy samples. I think in general this is used a lot in research and a policy that is fairly widespread in terms of other types of studies in which people participate. People may choose to change their mind and not stay in the study but I think the usual assumption is that they won't be providing further information to be used for further analysis but not that you destroy what you have already collected (#59 US, medicine, bioethics, government, R/A/S).

Arguments against policy (a)   Most interviewees who ruled out option (a) argued that it is "unethical" (#41 US), "immoral" (#16 US) or "absolutely contrary to ethical rules" governing human experimentation (#09 US).[4] These rules, as understood by these respondents, plainly include the right to withdraw: "Patients have always the right to withdraw" (#57 IS) and "[Policy (a)] violates the fundamental right to withdrawal" (#39 US).[5]

---

4 #41 US, bioethics, law, university hospital, R/A; #16 US, philosophy, bioethics, university, N, #09 US philosopher, catholic theology, university, N.

5 #57 IS, No. Am./Europe, medicine, university, S; #39 US, Europe, philosophy, bioethics, government/university, R/A.

Respondents underlined that the right to withdrawal is "important" (#48 IS, No. Am./Europe, philosophy, university, R/A) and that it includes not only the right to abstain from physical presence but also the right to put an end to the use of one's human biological material.

> The one that is totally unacceptable is (a). ... Every time you do a research study the participant has the right to withdraw whether it is a body part or their physical presence. So (a) is unacceptable (#64 US, medicine, bioethics, university hospital, N).

> You cannot accept it [option (a)]; donors must be given always the opportunity to withdraw and to ask that their samples are not used (#31 IS, Europe, medicine, genetics, university, biobank, S).

In line with these statements, option (a) is considered incompatible with the definition of "withdrawal" in research ethics.

> I think (a) is really bad because it just does not allow people to withdraw their sample and I see withdrawal as a part of a participant's rights in research and participation. To me this is unacceptable. What has been used will remain and their sample will remain physically and it can be retested, I think this is unacceptable (#39 US, Europe, philosophy, bioethics, government/university, R/A).

> In the fact you are saying you are redefining the rule of withdrawal ... if you offer them a right of withdrawal then it seems to me once they exercise it you have to take this seriously; you cannot proceed [and use samples] as you wanted before (#02 US, philosophy, catholic theology, university, N).

Respondents against option (a) referred to the principle of respect for persons who participate in research (#65 IS, So. Am., life sciences, genetics, university, A). This principle is seen to imply, first, that research participants' wishes trump other values such as the benefit for science or society and, second, that research participants do not need to provide a reason for their withdrawal.

> In my definition of a lexicon of ethical principles the experimental subjects' wishes trump other things, and if they got into it and [then] want to change this for whatever reason, they should get out of it (#09 US philosopher, catholic theology, university, N).

Interestingly, this commitment to the value of autonomy is shared by most researchers among the respondents. Although they acknowledged that option (a) would be easier for scientists, they judge the associated violation of research subjects' rights as being so severe that it could never be justified.

> I don't like [option] (a). I understand the burden to the researcher to keep track of all this stuff, but I really think that the participant should have the right to say: you can't use my samples anymore you can't use my data anymore ... patients may be upset by the way the process is going and don't want to participate and their lives have changed and to say to them once you give this you can't ever take it back ... I would not like that as much but I understand it is easier for the researcher. But I don't know that that should be the

motivating force. ... I guess we would not use (a) ever (#96 US, genetics, medicine, social sciences, university hospital, R/A/S).

Interviewees think that policy (a) "puts unnecessary pressure on the participants to make a decision at the beginning of the study whether actually to give or not to give their samples" (#12 US, Africa, genetics, university, R/A/S). Respect for autonomy means that research subjects have the right to change their minds (#47 US, law, social science, government, A). Even if they agreed to participation when they provided informed consent, they might have misunderstood parts of the consent or new elements might make them reconsider their consent.

> It is somehow authoritarian; persons always have a right to change their minds. Alternative (a) would be too strong (#79 IS, So. Am., medicine, university, S/A).

> Informed consent has to allow for persons to change their mind, and [policy] (b) does not allow it and [policy] (a) [is] similar. If (a) and (b) are in the informed consent, it is still not acceptable because that presumes of course two things, one that it is perfect informed consent which never ever does occur—there is no such thing as perfect informed consent—and secondly, that things cannot occur temporarily that change a situation whereby a person could say: 'under these conditions certainly I would consent; under these, no' (#80 US, genetics, bioethics, philosophy, theology, university, R/A/S).

Several interviewees used a prison metaphor indicating that so-called Ulysses contracts are not ethical. Option (a) is "like signing yourself up for prison" (#55 US, philosophy, bioethics, university, N). "I would reject [option] (a) ... We don't lock in people" (#62 IS, Europe and No. Am/Europe, medicine, law, university, R). Yet not all respondents who are against policy (a) share the view that Ulysses contracts are unacceptable. One respondent opposed option (a) because he does not find it "fair" to participants; however, he explained that it would still be ethically acceptable.

> [Option] (a) is, I don't think, fair. This is a linked sample: it is not impossible to remove people. ... [Option] (a) I think is feasible, it is workable. Ethically I don't see why not. People basically said you can take my blood ... and if I stop participating, to stop participating means that I stop participating going forward. ... Now some people would say because your stuff is still out there you can stop using it, but you know our stuff is out there all the time, we have given stuff for clinical purposes, medical information is taken, we can't stop the use of that stuff. So here, as long as we tell people about that, I think it would be OK to do it (#10 US, humanities, law, university, R/A).

Furthermore, some interviewees opposed option (a) because it implies insufficient protection of confidentiality (#46 IS, No. Am., law, social science, NGO, A).

> I can see [the advantages] with [options] (a) and (b) making material available. It looks as if (b) provides better protection for confidentiality, so I suppose (b) is to be preferred [over option (a)] (#02 US, philosophy, catholic theology, university, N).

Respondents from the international sample ruled out option (a) also because they think it would overall not be beneficial for research. Society would lose trust and it could become more difficult to recruit research subjects if such restricted withdrawal policy is used.

> I certainly say that you cannot have a system where you have no right to withdraw because people would very quickly lose confidence in that system and they would refuse to participate (#82 IS, Europe, natural science, A/R).

> [Option (a)] would be probably the best for the researchers, the most secure. The samples are there, so there is no more discussion about it. I wonder if you have more difficulty in recruiting if you tell people 'look you gave us your sample and you won't be able to get any information back and we can use it for whatever we want' ... because genetic information can change people's lives. It may be difficult to recruit (#60 IS, So. Am, pharmacology, medicine, government, S/R).

Another participant dislikes option (a) because it is less easy to put into practice for researchers than might be assumed.

> [Option] (a) may be more complicated than it may seem because the participants come into the project together with some blood relatives. If some but not all blood relatives want to withdraw, it is going to impact. Those who want to remain may not be able to facilitate scientific research on the disease (#13a IS, No. Am./Asia, philosophy, university, A/R).

One participant argues that the justification for withdrawal policies depends on the purpose of future studies. The more general the consent and the broader the range of studies, the more complete the right to withdrawal needs to be. Policy (a) would not be sufficient for a study in which future purposes were defined broadly.

> [I rule out option (a)] depending on what the agreements were about using the specimens in the first place. If it was a wide-open agreement to use them for whatever you wanted, I would have a lot more concern about it (#18 US, genetics, NGO, R/A/S).

Finally, it could be possible that respondents against option (a) assume that withdrawal of consent by research participants is a rare event and that therefore option (a) is not needed, as is suggested by the respondent who stated "They [research participants] should have the autonomy to sometimes remove their information" (#65 IS, So. Am., life sciences, genetics, university, A).

Arguments in favor of policy (b): irreversible anonymization in the case of withdrawal Respondents in favor of option (b) argued that "destroying them is not acceptable" (#06 IS, So. Am, philosophy, bioethics, university, A) because it has negative consequences for the research and would, in addition, not be in the interest of those who want to remain in the study and who would have made efforts or taken risks in vain.

> If many patients withdraw their consent from the study, and samples and data are destroyed, the study may lose its value. Even those who have donated samples and want to stay in the study will find themselves being part of an underpowered study (#56 IS, Europe/Europe and Africa, medicine, life sciences, government, hospital, S/A/R).

Similar to the respondents who favor option (a), many respondents who prefer option (b) think that "it must be included in the informed consent" (#17 IS, Europe, law, university R/A). If this is the case, interviewees consider option (b) the best balance between respect for participants" autonomy rights, protection of research participants' confidentiality (#02 US, philosophy, catholic theology, university, N) and convenience for the researchers because they maintain the possibility to continue studying the samples after they have been anonymized.

> It [option (b)] is the most convenient, an intermediate between protection and the ability to recruit (#60 IS, So. Am, pharmacology, medicine, government, S/R).

> I am inclined to choose (b) on the grounds that by anonymizing it—let's say—it is really anonymized, I am taking irreversibly [and] seriously here... the participant really has no grounds for complaint because nothing can be linked to them, but it does allow the researchers to take advantage of the information that people have already agreed to be used. So, without having given it any further reflection that is the one I would gravitate towards (#45 US, bioethics, philosophy, government, O).

> People make a big thing of that [withdrawal]. I am not that upset about it, technically. ... The doctor could say: we keep your sample, but we destroy the link. So all [we] know [is] that you were in the study at some time but there would be no way to know which your sample is and link it back to you except through statistical matching or something (#10 US, humanities, law, university, R/A).

Thinking also of the consent process, one interviewee believes that anonymization is an acceptable form of withdrawal, in his personal experience, at least for research participants who agree with policy (b):

> I am thinking of people that I have talked to about samples, and some of them don't mind as long as their names are not attached. Therefore for them (b) would be fine. They will just agree that they are anonymized (#37 IS, Oceania, social sciences, bioethics, university, A).

One respondent argued that research participants should not have the right to ask for destruction of their samples because their wishes are often not their own but influenced by the "political and financial agendas" of others. He does not see any justification to let these influences hamper research.

> I don't know whether this is politically correct or not, but I would be completely against [options] (c) and (d), precisely because of the HGDP [Human Genome Diversity Project] experience. ... There is the risk that sometimes political and financial agendas intervene, and indigenous patients and groups may be motivated [to withdraw] by NGOs that have political interests. It has certainly happened with the Human Genetic Diversity Project. I am OK with that [option (b)]: if someone desires to withdraw consent, there is irreversible anonymization. ... I am not happy with destroying samples and/or data. ... I would favor (b): they [samples] are irreversibly anonymized but they stay in the bank (#56 IS, Europe/ Europe and Africa, medicine, life sciences, government, hospital, S/A/R).

A related concern was raised by a respondent from Asia who explained how important it is to use only withdrawal policy (b). Withdrawal should always mean anonymization, otherwise "nobody would make any investment' because it would be possible to make companies collapse by paying research participants to withdraw their material from the biobank.

> Somebody would say 'I've gone so far in the research and now somebody wants to withdraw, where do I go?' [Under policy (b)] whatever investigation has been done so far, [samples] can be anonymized and still be used. I think this is the most appropriate way of utilizing the investment that you made. Otherwise, what would happen is that nobody would make any investment—you are pursuing a biobank to have data to make a discovery, and then you find there are 100 [participants from a] Jewish or Amish population involved, and I somehow convinced them by paying them money or whatever that is, and they should withdraw, and thereby I can [make] collapse the entire structure of a company. And that's not the right thing to do (#78 IS, Asia, life sciences, government, S/A).

Interviewees in favor of policy (b) disagreed about its ethical and legal premises. One respondent who thinks it is legal referred to the fact that samples are not owned, but that the biobank becomes the custodian of them. He is convinced that custodianship includes the right to continued use in anonymized form.

> As a custodian, you can keep the patients' names and the data anonymized, so this is very important. If they are anonymized, then it is alright, you can still keep it (#73 IS, No. Am./Asia, life science, government, S).

Another respondent fears that ethics boards will not accept the withdrawal policy (b) which he prefers himself.

> Here the samples are anonymized once the participant exercises the right of withdrawal, so basically here is no right of withdrawal anymore. For some Ethics Boards this is not going to be acceptable ... I think the current Ethics Boards would favor [policies] (c) or (d) (#60 IS, So. Am, pharmacology, medicine, government, S/R).

Several respondents find destruction ethically more correct, but chose policy (b) because destruction is often impractical according to their experience.

> Ideally, I like option (d); however, when I look at the realistic situation—for example, some institutions make purified DNA from different participants to promote the efficiency of finding polymorphism—when you use such method to purify DNA, it is impossible for participants to withdraw their sample. Therefore, considering the present situation, I would like to choose option (b) (#94 IS, Asia, life sciences, government, A).

One respondent, although in favor of policy (b) explained that anonymization is not sufficient to protect participants' privacy. He finds therefore policy (b) easier acceptable for information than for samples.

> It [option (b)] allows one to retain the information. It is unlinked from the individuals so it shouldn't be able to compromise them in any way if it is only data. Continue using samples would bother me a little bit more but I suppose my problems lie in the nature of

the colorectal disorders being evaluated. The rarer the condition, the more likely one is able to identify the individual independently. … We do this in newborn screening. We get one [name of a rare disorder] in three states in a year if we are lucky. I would always know any data ever published on the disorder if I manage the patient, I would always know who they were (#18 US, genetics, NGO, R/A/S).

Arguments in favor of a combination of policies (b) and (c)   Some respondents proposed combining policies (b) and (c) to resolve the dilemma between ethical correctness and practicability or reasonableness. One participant gave an interesting description of two situations, involving tissue arrays and mixed samples, in which destruction of one single sample is impossible and in which the only way to do it would be "unreasonable" because it would imply the destruction of thousands of samples from other participants.

> We recommend that they [samples] are actually destroyed but there are … situations in which this is unreasonable. The most obvious situation is the so called 'tissue arrays.' The principle is that it is a sort of miniature sample, typically of tumors—this is the most popular. What happens is that a little robot is retrieving very small samples of cancers and it is putting up to 2,000 cancer samples on the same little slide. And these are small slides…this is so to say a big revolution in the analysis of biobank specimens. The specimens can be put on such arrays. However, if the person withdraws and the sample has been put on the array, in that instance it cannot physically be withdrawn without also destroying the other 1,999 subjects. There are [other instances] where blood samples from several persons have been mixed and the part from one person cannot be identified, and withdrawing would mean destroying the whole [tube]. We strongly recommend that the sample is actually destroyed, but in the situations that I have just described, where it would be unreasonable to destroy because it would mean to destroy other people's samples too, we are strongly in favor of research and for the samples to be preserved. Under these circumstances, we think that complete anonymization should be allowed (#32 IS, Europe, medicine, university, S/A).

While the respondent quoted above supports a variation in which policy (b) is only used if (c) is "unreasonable," other interviewees think that research subjects should have the right to choose between both. Although the logical way of implementing a right of withdrawal is destruction of the samples, research participants' autonomy rights are sufficiently respected if they are informed about a policy which restricts withdrawal to anonymization. But, in the view of these respondents, if participants really want to withdraw their samples, the samples should be destroyed.

> I don't think any of the four is entirely satisfactory because it depends how you look at it. Excluding permanently all the information after a participant withdraws his or her consent is a logical option because if there is no informed consent of the patient, you cannot use the samples. [Alternatively], there could possibly be two options, (b) and (d). [Option (b) is OK] provided what I said earlier, that is that the participant knows that the material will be anonymized and used. And if the patient withdraws the consent, information and sample must be removed from the repository and destroyed. And this is a logical way to deal [with withdrawal] when the person doesn't want his/her sample to be used (#86 IS, So. Am., medicine, university, A).

Finally, one participant proposed to allow the destruction of samples, but to keep the information. He considers this to be the best way to preserve anonymized genetic information for the public good, while respecting participants' rights to withdraw their samples.

> By combination of [options] (b) and (c), what I meant is that: ... If a patient requests that the sample is destroyed, that seems hard not to follow. So the information that has gone to the project, I will not allow that to be removed, once it has been anonymized and unlinked. For me it is very easy to distinguish the information. Human genome information is [a] public good. If a cell was taken from a patient, it is a cell, there is no way around it [the cell can be destroyed], but when I talk about common good I talk about the information not the specimen (#62 IS, Europe and No. Am/Europe, medicine, law, university, R).

Arguments against (b)   Several arguments raised in the interviews against policy (a) were also used against policy (b). These objections were sometimes framed in terms of principles (such as a person having a right to have samples withdrawn from the repository) and sometimes in consequentialist terms (such as protecting against the harm if a person could be linked to the results of a subsequent genetic analysis of their sample still held in a repository after they had "withdrawn" from the research). The first view is reflected, for example, in the response that "informed consent has to allow for persons to change their mind and [policy] (b) does not allow it" (#80 US, genetics, bioethics, philosophy, theology, university, R/A/S), while the second was heard from respondents who argued that policy (b) is simply not equivalent to withdrawing samples from the biobank because the right of withdrawal means that the samples should never be used (and put the participant at risk) again.

> The problem with [option] (b) is that their samples would still be used as with [option] (a) (#71 US, Europe, law, philosophy, bioethics, university, N).

> [Option] (b) is not a withdrawal. It is an anonymity guarantee (#09 US philosopher, catholic theology, university, N).

Even respondents who hold a strong rights-oriented view were not hostile to the research. As explained by one respondent from the US, researchers have the right to persuade participants to stay in the study, but not to continue using their samples after withdrawal.

> [Options (a) and (b)] should be ruled out because I don't think they allow people to have their rights respected. I think if people want to withdraw their stuff, they want to withdraw their stuff. I would try to persuade them against it. I would suggest what safeguards are out there so that they won't be harmed, but I would not prohibit them (#33 US, bioethics, medicine, government, A).

Respondents who oppose policy (b) do so to varying degrees. Some find destruction more appropriate, but think also that policy (b) "is not ... unlawful" and that "in the end one could come away with it" as the policy for the biobank (#53 IS, Europe, medicine, public health, Int. Org., R/A/S). Others reacted against policy (b) almost as strongly as the firm opposition to policy (a):

I think [policy] (b) is outrageous. If people have a right to withdraw it means withdraw. It does not mean that the repository has the right to anonymize them and to continue to use them. And I feel not quite the same way about policy (a) but close. If the right to withdraw means anything it's got to be the right to say that you can't use this information and these samples anymore, period (#92 US, law, medicine, genetics, university, R/A).

Interviewees who take a consequentialistic approach to the right of withdrawal hold different views concerning whether irreversible anonymization offers sufficient protection to research participants who wish to withdraw. One respondent from North America considers it "not enough protection" (#46 IS, No. Am., law, social science, NGO, A). Other interviewees from North America and Europe think that policy (b) offers better protection of research participants than policy (a). They favor a more extensive withdrawal policy than (b) for a different, rights-based reason: option (b) does not respect sufficiently the human right to abstain from research which seems unacceptable to a particular research participant.

[Option (b), irreversible anonymization,] is ok on the risk and benefit view but it is not on the human rights view. If the person says, 'I understand data are being used to study abortion or depression or something like that, and I don't think they should be used for that' then to continue to use them [is unethical] (#68 IS, No. Am./Europe (nationality: No. Am.), life sciences, consultancy R/A).

Further practical arguments raised against policy (b) were that research participants would not agree to option (b) and that this option is not in the interests of research subjects because feedback is not possible (because of the anonymization).

[Option] (b) is simply unacceptable unless that's the deal with the participants, so that rather than withdrawing there will be anonymization. It is a question of negotiation but I would not rule it out that participants would agree to this one (#48 IS, No. Am./Europe, philosophy, university, R/A).

The advantage of going back to the source would be lost (#42 IS, Asia and No. Am./Asia, medicine, bioethics, government, A).

Finally, the right of withdrawal was explicitly linked to the underlying requirement of consent for research uses of stored samples by one participant, a member of a research ethics committee, who would refuse protocols based on policy (b).

We have [examined] some protocols offering this [withdrawal in form of anonymization]. I represent the lay persons in the ERC, and my view is that whatever samples are used, from a participant who has consented to certain research, ... should never, never be used in other research (#22 IS, Africa, science, self-employed, A).

Arguments in favor of (c): withdrawal implies destruction of all samples and information stored in the biobank   Respondents in favor of policy (c) disagree with those in favor of policies (a) and (b) about the meaning of the right of withdrawal. For most interviewees who favor policy (c), this right encompasses being able to demand destruction of all samples and information stored in the biobank.

[Option] (c) is best: when participants withdraw consent samples are removed from the bank. This basically respects people's rights of withdrawal. They contributed to this thing and now they are withdrawing from this thing (#10 US, humanities, law, university, R/A).

Samples should be destroyed (#77 IS, Africa, social sciences, NGO, A).

If the person doesn't want to cooperate, you have to destroy the samples (#79 IS, So. Am., medicine, university, S/A).

Either [option] (c) or (d) because I think a right to withdraw means something like that (#95 US, bioethics, philosophy, theology, university, N).

The right of destruction of samples was linked by one respondent to the position, to which the respondent also subscribed, that research subjects are the owners of their samples:

I like [option] (c) the best because I think that the sample is in the ownership of the patients so if they want to withdraw consent at that point their samples should be removed from the biobank and destroyed. They should not continue to be used (#64 US, medicine, bioethics, university hospital, N).

Yet not all respondents who favor policy (c) share the view that destruction of samples represents the correct way to interpret the ethical right to withdraw from research.

This is a question that is really complicated because in research there is generally understood to be a right to withdraw from research but that really applies to having a physical active engagement with your life or body in the research and that you don't have to have then anything to yourself anymore, whereas the sample it is not the real person anymore so I don't know whether the right to withdraw should apply in this case (#26 US, medicine, social sciences, government, R/A).

Not being sure about what the right to withdraw encompasses, this respondent chose option (c) because he regarded it, in contrast to option (d), as more practical:

I think (d) is not a reasonable option and people have to be informed that once they do [participate] these things will be shared … that the decision they make will at least be partly irreversible … If only very few people withdraw, [option] (d) could be acceptable. But then [option] (c) would still be fine. It is just difficult to imagine that someone becomes so upset that they want to withdraw from every ongoing research project that one needs [option] (d) (#26 US, see above).

For some respondents, policy (c) is mandatory. According to a researcher from Europe, the view that withdrawal must include destruction is absolute in his country.

From the date when withdrawal is known no further analyses on identifiable samples are allowed. This is something that, at least in our country, is considered absolute. I know that there are other views that participants may not exercise their right under (a) here, but it is very rare, and there is nobody actually working with biobanks who doesn't grant the right of withdrawal, and no further analysis will be done after that point (#32 IS, Europe, medicine, university, S/A).

Likewise, others judge option (c) as the necessary default position; that is, its supporters regard it as the only acceptable version of the right of withdrawal if the consent form did not set forth an alternative withdrawal policy.

> I think if the patients are told when they grant consent that there are limits on their ability to withdraw, so when they consent to [policy] (a) then it's fine or [policies] (b) or (c) or (d). If they are not told, then option (c) is the best (#71 US, Europe, law, philosophy, bioethics, university, N).

Others saw policy (c) as the natural manifestation of the appropriate stance regarding consent in all research involving biobanks. In this view, one-time consent cannot be binding forever, as would occur with policies (a) and (b), because when samples are stored in biobanks, participants cannot be provided with complete information about as yet undefined future studies.

> [Option] (c) [is the best policy] because I think we have to continue to give people the option to get the information out of this back because we are not quite certain what scientists could use this information for (#12 US, Africa, genetics, university, R/A/S).

Many respondents did not view option (c) as mandatory but simply chose it on the ground that it is preferable to all the alternatives, which they find too extreme.

> Option (c) is the best. The problem with [option] (b) is that their samples would still be used as with [option] (a). [Option] (d) goes a little too far, in that it requires researchers to destroy information that has been gleaned with consent. If a patient has consented to research it is too late to retrieve consent so [option] (c) strikes the right balance (#71 US, Europe, law, philosophy, bioethics, university, N).

These interviewees saw policy (c) as offering the best balance between feasibility and respect for research participants' autonomy.

> I sort of feel like it's [option] (a) or (c), it depends on what they consented to at the outset. I think [option] (b) is a bad choice, it's the lower end of my preferences, [while option] (d) is most respectful in some ways of the participants but most burdensome to researchers so I think there is a feasibility concern. For practical reasons I think (d) is not a good choice (#07 US, social sciences, bioethics, university, O).

Thus, policy (c) is considered an adequate way to grant participants the right to choose while being the best compromise in terms of costs for the researcher, especially compared to policy (d) which is judged too harmful for scientists.

> I am most inclined to favor [policy] (c). I think this is a matter of affirming the liberty of the participant in the project. I realize this produces some cost certainly for the researchers and in practice the best would be to try to persuade participants not to withdraw, but it ultimately is a matter of their consent being decisive. And it seems to me that the first two [option (a) and (b)] don't really acknowledge this … [Option (d) goes too far]. If the initial

consent was properly explained then the participant even if he changes his mind has no ground for complaint if the material has been used and conclusions have been drawn from it (#02 US, philosophy, catholic theology, university, N).

Yet several others who regard (c) as best option would impose heavy restrictions. One respondent explained that option (c) is adequate, but underlined that "you have to have very strong reasons, ethical or commercial, in case of dispute so that you can [withdraw your consent]. Withdrawal and destruction of samples can only be done in certain cases" (#61 IS, Africa and Europe/Africa, life sciences, university, R/S/A).

Another respondent said that samples should be destroyed only if certain conditions are fulfilled, such as when there is a strong basis for thinking that retention of the samples would risk harming a participant, such as when his or her identity has been exposed.

> It [option (c)] would be appropriate only if the [identity] of the participants is somehow revealed, and there should be proof about it, not only some doubts. The sample should be removed and destroyed *only if there is a good reason to do so*, and it has to be demonstrated (#24 IS, Europe, medicine, consultancy, A/R).[6]

He justified his position referring to the public interest in samples used for research:

> I am not in favor that when someone withdraws, you remove and destroy the sample. You have to keep public interest in mind. What I would suggest is that you add something at the beginning that allows the samples all the way to be analyzed. I would like an option where withdrawal and destruction of samples can only be done in certain cases, when the subject or the donor will request so … you [should] indicate at the beginning in your consent form that the samples will be analyzed and you make restrictions on withdrawal. That would be better because, once you donate the samples they are meant to be analyzed (#24 IS, Europe, medicine, consultancy, A/R).

Finally, one respondent from South America chose policy (c) because he finds destruction might be easier than irreversible anonymization.

> This is the most difficult question … It is also difficult to understand the right of withdrawal. It is much easier in clinical trials when they are testing a new drug. [Anonymization is difficult because] then we have to anonymize totally the sample in all different biobanks. If somebody asks for it, the sample shall be destroyed (#49 IS, So. Am., biology, bioethics, university, A/R).

Arguments against policy (c) Different criticisms of policy (c) were expressed by respondents favoring policies (a) or (b) than by those who prefer policy (d). Respondents in the former group tend to regard policies (c) and (d) as logistically difficult and too burdensome on researchers (#36 US, medicine, genetics, government, R/A/S). The scientists among them who had used or stored samples described option (c), together with (b) and (d), as "headaches" (#98 US) and would

---

6 Emphasis added.

rule out all three "for practical reasons" (#97 US).[7] One interviewee described in more detail why, in his experience, policy (c) is hampering research; although he would like to use a policy such as (c) for ethical reasons, he considers the practical problems prohibitive:

> Destroying the samples, the thing about that [option (c)] is that it does cripple or hamper the development of the research because so many studies in genetics—for them to be effective—demand restudy upon restudy upon restudy ... What we are learning is that the first five studies are done in such a way that they have to be redone in a slightly different way and you sort of learn as you go, so if you kept changing which samples were available to you, that would be problematic in a way that is probably not necessary. I am sort of torn, really truly: if someone wanted to pull their sample out they should be allowed to pull their sample out, but I think it is so impractical. How do you replicate the DNA, how are you sharing it? [Whereas option] (B) you could do (#52 US, social science, bioethics, university, R/A).

Policy (c) is most frequently not rejected in an outright way, but rather is judged to have too negative consequences for science without providing any advantage to research subjects in terms of protection, as compared with the other withdrawal policies.

> I can live with [options] (c) and (d) but they seem to be more either onerous or place too great a penalty on the science without yielding any really enhanced protection of subjects over [option] (b) so I think they are not preferable (#45 US, bioethics, philosophy, government, O).

Those who favor a more extensive withdrawal policy than (c) criticized that it does not sufficiently respect research participants' autonomy rights which include the right to destruction of all of their samples and information. Option (c) is said not to "satisfy the ...participants' ability to withdraw ... At least in many circumstances, we have to go to [option] (d) to be effective, [and option] (c) is only a very partial approach" (#29 US, law, government/university, O).

Another respondent explained that his objections to policy (c) are similar to those to policies (a) and (b). He thinks that policies (a), (b) and (c) are ethically problematic because he believes that if explained correctly, all rational persons would reject Ulysses contracts which take the form of incomplete withdrawal policies.

> I don't object to people being offered that arrangement. I think it would be poorly advised for them to agree. If worded properly what you would have to say to the donor is: 'once you agree to contribute we have a unilateral right to use this information in any way we want, including ways we have not thought up yet and that will include ways that predictably you might have violent objections to. Do you consent?' And I think rational people would say 'no' to anything but option (d). ... I think there are ethical problems with offering only any of the other options which in effect are offers for Ulysses contracts where consent cannot be withdrawn. I don't oppose Ulysses contracts as such but I think that most people if they understood the deal ... most reasonable people would decline the offer which means the

---

7 #98 US, philosophy, medicine, genetics, university, R/A/S; #97 US, medicine, genetics, university hospital, R/A/S.

databank would be [limited] by having only sources from unusually compliant, passive people who did not worry about their rights (#38 US, bioethics, philosophy, theology, medicine, R/A).

Arguments in favor of policy (d)   Most respondents favoring policy (d) seem to regard it as the only meaningful way to respect the right to withdrawal.

> When people agree to participate in a repository of this nature then if the right to withdrawal is going to mean anything it has to mean the ability to withdraw the samples and medical information that has not already been analyzed (#92 US, law, medicine, genetics, university, R/A).

Policy (d) was favored for a variety of reasons, from the practical to the principled. For example, some respondents called it the "cleanest" (#04 US) and the most "logical" (#34 IS and #93 IS).[8] These views seem to reflect assumptions about the underlying meaning of the right of withdrawal. For these respondents, this right could not mean anything other than destruction of all existing samples. Some insisted that anonymization would not be accepted in their country as a form of withdrawal (#78 IS, Asia, life sciences, government, S/A). Policy (d) is preferred because biobanks should not be an exception to what respondents see as the usual right to withdrawal in research ethics.

> I would go with (d). Presumably unless someone gives me some compelling reason why not, generally speaking people are free to withdraw from participation in research should they change their mind about the risks or benefits or develop an objection to the research itself. I don't see any reason why this should be an exception (#44 US, bioethics, social sciences, A).

As was true regarding the other options, respondents favor option (d) on two quite different grounds. Some see it in rights terms; they give precedence to research subjects' autonomy rights whatever the consequences for research. Others would choose policy (d) only if the negative consequences for research don't exceed certain limits.

Respondents in the first group tend to refer to fundamental elements of classical research ethics such as the principle that the rights and wellbeing of patients involved in research always take precedence over society's interests.

> Patients' interests should override society's interests. It is not necessary that the patients give up their control over their body parts (#11 US, medicine, bioethics, university, R/A).

Basically, respondents in this group regard respect for persons as the overriding principle.

> For me, I noted that it is option (d); I think it is logical because the respect of the participant is a priority in this aspect (#93 IS, Europe/Middle East, biology, genetics, biobank, S).

---

8 #04 US, medicine, genetics, university hospital, R/S; #34 IS, Europe, medicine, genetics, biobank, S; #93 IS, Europe/Middle East, biology, genetics, biobank, S.

If you believe in respecting persons' rights then you would be the most conservative and you would say [option] (d) (#01 US, medicine, bioethics, university, N).

The goal to achieve by withdrawing consent is that research participants decide actually that they don't want their DNA used and the associated information…I think that should be respected (#72 IS, Europe/Africa and Europe [nationality: Europe], medicine, S/A).

The main advantage attributed to policy (d) was that, in the words of a North American lawyer working with an NGO, "it provides the most protection to the individuals to exercise their right" (#46 IS, No. Am., law, social science, NGO, A), a view shared by participants from different regions, such as this Asian government physician working in bioethics:

I agree with part (d)…[It] provides protection to the right of the participants, who at a later date might withdraw (#42 IS, Asia and No. Am./Asia, medicine, bioethics, government, A).

For respondents who favor option (d) on such grounds, the right is not subject to limitation simply because some think is it being exercised unwisely. For example, one respondent from the US not directly involved in biobanks research argued that respect of autonomy implies respect for reasons that are motivated by fears even if these fears are not well grounded.

I would put the strongest standards of permissible withdrawing on to such samples simply because in genetics I can understand people's fears of it being out there even if they are not particularly well grounded. So I would put high priority on the participants' own subjective sense of the importance of the specimen and the knowledge (#16 US, philosophy, bioethics, university, N).

These respondents insisted that patients don't have to provide a reason for withdrawal and that the right to withdrawal is absolute even if withdrawal is disruptive for science.

I think participants have their own reasons whatever they are to withdraw consent and participants should be able to withdraw at any stage however disruptive it might be in the science and that is an unqualified right and their withdrawing of consent should be such that samples are destroyed and no data collected could be used that are not already submitted for publication. I think they have an absolute right all the way through (#20 US, law, philosophy, university, R/A).

Indeed, one respondent insisted that the possible negative consequences for research from subjects withdrawing samples are simply a risk that researchers have to shoulder.

Sometimes if you are doing clinical research…[in a small study] if enough people withdraw from the protocol in the middle of the trial that is the end of your trial. You are dead but that is part of doing research and I don't think it is any different for researchers in a biobank (#41 US, bioethics, law, university hospital, R/A).

Yet a sizeable percentage of the respondents who favor policy (d) fell into the second group who would impose restrictions when they thought this type of withdrawal would unduly interfere with scientific quality. For these respondents, the number of research participants who are withdrawing becomes an important criterion by which to judge whether the policy will be too harmful for the research or not.

> I see that this [option (d)] can be difficult for the research and I am not sure whether it [data and samples] even can be retrieved from the investigator. If it can it ought to be, if it can without totally destroying the whole experiment but I think all of this should be made clear at the front end to the participants. If it cannot be retrieved or withdrawn from the researchers in the fiel,d that should be known by them. All the possibilities of withdrawal and the impossibilities should be made clear to them in the information ... [Option] (d) is most preferable if it is possible without messing up the experiments (#09 US philosopher, catholic theology, university, N).

If the number of participants who withdraw is too high, these respondents would prefer option (b):

> It depends on the example, here colon cancer. If you have got 3000 and 5 withdraw this is not going to affect the scientific integrity of the study then I would want something like (d), do everything. It is not much scientific cost that is what the participants want. But if it is going to undermine the study I would want something like (b) ... The threshold? Talk to a statistician, and people who have experience with this kind of study (#27 US, philosophy, bioethics, government, R/A).

Therefore, according to another respondent from the US, before deciding on a policy regarding withdrawal, one needs first to evaluate how frequent withdrawal is. Since the respondent thought it is probably infrequent according to empirical data, researchers could adopt policy (d).

> First of all you have to make an assessment a priori how frequent this is likely to be, and this is likely to be very infrequent. As long as it is likely to be infrequent, I think you can ask the researchers to do more so I would say [option] (d), and I don't think it's going to be burdensome because all the data suggest that this is not going to be a very common event (#33 US, bioethics, medicine, government, A).

Indeed, a number of interviewees favoring policy (d) had already made particular assumptions about the answers to several empirical questions. They think that research participants who wish to withdraw do not want anonymization but destruction of all of their samples. They believe also that policy (d) can be realized without significant costs to the research especially because withdrawal is a rare event.

> I think it [withdrawal under policy (d)] is rarely going to happen, but it ensures maximum concern for the participant's wishes. I think it is more a theoretical provision than anything that would happen, anything more than rarely ... but if someone passionately wants to withdraw then they want to withdraw. They do not want you to pretend that they have withdrawn and only anonymize the samples; the material has to be withdrawn completely. It can be done at low cost. If you run your study and after 5 years 20% want to withdraw then you have to redesign the study, but we have no evidence right now especially if the

double coding reassures people … At this point you should go with [policy] (d) (#55 US, philosophy, bioethics, university, N).

As with the other options, some preferred option (d) for consequentialistic rather than purely rights-based reasons. Several interviewees favor policy (d) because they think that it is most protective of research participants, while deleterious consequences are unlikely.

> I vote for [option] (d). I just think it is the most protective, and I think in fact it is unlikely to have much in the way of deleterious consequences because I would doubt that a lot of people would withdraw. Now, that is an empirical question. I have no basis for this. But if people decide to volunteer for research and if they choose to withdraw they should also be allowed to withdraw (#25 US, philosophy, government, N).

> Among the four options that are offered, I think (d) is the most protective one …It should be feasible in research (#35 IS, Europe/So. Am., medicine, genetics, university, S).

One practical issue generated considerable discussion among respondents who favor option (d), namely the point during research up to which withdrawal should be allowed. Several think it is acceptable to permit withdrawal until the submission of research results for publication. One respondent said that "things can languish in final datasets for a long time sometimes, [but] once it is submitted [for publication] it is too late" to withdraw the data (#04 US, medicine, genetics, university hospital, R/S). Others argued that the endpoint should be earlier. One interviewee favors the moment when data are generated because they cannot be "un-used":

> [D]ata that has been already generated shouldn't be able to be withdrawn. Once the information is there and available—it has been used—you can't 'un-use' it … So essentially you need to be able to draw a line and say 'my participation stops here and there is nothing in the future that you can do with it' but there is not a retrospective right of withdrawal or un-participating for things that have already happened (#82 IS, Europe, natural science, A/R).

In addition, respondents argued that a late endpoint would permit undue and "unfair" influence on studies by third parties.

> We could not go in the dataset, pull out the data and destroy them. This would not be fair because if someone doesn't like the results of the study he would go and ask for withdrawal of the sample … If we try to sort of remove data from the lab books, then we are half way to having irreproducible science … And it is really for the ethics to prevent scientists from erasing things from the original lab books … that should not be allowed under any circumstances (#32 IS, Europe, medicine, university, S/A).

However, other respondents questioned this approach by suggesting that, if a late withdrawal occurs, samples should not be used and that "one has to re-do the statistical analysis, because you did it with more cases than what you have in reality" (#53 IS),[9] and that, at any rate, "it is the obligation of the researcher to make the

---

9 #53 IS, Europe, medicine, public health, Int. Org., R/A/S.

research so interesting and solid that nobody wants to leave" (#68 IS, No. Am./ Europe (nationality: No. Am.), life sciences, consultancy R/A).

Arguments against policy (d) Respondents ruled out option (d) in the first place on practical grounds. They indicated that it is "impossible to realize" (#70 US)[10] "because the problem is that if you have already done research on the samples it can be very difficult to start removing them" (#99 US)[11]. In addition, they said that it is too difficult to contact all researchers who have received samples.

> For [option] (d) it becomes a practical question: how many researchers have used the data, how critical it is to their particular [project]. Once the data has been analyzed it is a little hard to remove them from research studies, so I think (c) is probably the best option … if researchers have already used these samples to generate some data I think the researchers should be allowed to go ahead and use that data (#23 US, medicine, humanities, genetics, university, R/A/S).

On ethical grounds, respondents criticized policy (d) as harmful to researchers. These respondents think that one should strike a balance and limit research participants' rights if the harm to the research is disproportionate compared to possible risks to the research participants, especially if double coding is used, as in scenario A.

> Here again the principle is the balance between participants' rights and benefits of research and I think option (d) is the second [worst] because it goes to an extreme on the other side. So I would rank it as the second worst, because the researchers that have the samples do not have identifying information and they already use them and the data has already been processed as part of the research so I do not think it is a good idea to require them to destroy the samples and remove this data from the project (#39 US, Europe, philosophy, bioethics, government/university, R/A).

Most interviewees who spoke against policy (d) rejected it because of its consequences; they considered it acceptable if the withdrawal of samples and information does not alter the study. However, it is non-acceptable if important results have been generated from the material the participant wants to withdraw because in that case, according to a respondent from South America, the public value of knowledge should prevail over research participants wishes.

> [I]f the results are already there and are important, [samples] should be used (#79 IS, So. Am., medicine, university, S/A).

Respondents criticized policy (d) also because it might have other bad consequences forcing researchers to work in a hastened way in order to be protected against a possible withdrawal in the future.

> So if you say that the researcher is going to push to submit … early that is not necessarily a good thing (#80 US, genetics, bioethics, philosophy, theology, university, R/A/S).

---

10 #70 US, genetics, social science, bioethics, university, R/A/S.
11 #99 US, medicine, genetics, university hospital, R/S.

Moreover, interviewees who had used samples argued that although research participants have the right to control the use of their samples, they do not own the knowledge generated from the samples. Hence, they should not be allowed to withdraw research data.

> I did not like [option] (d) as much because the knowledge that has already come from their samples—they have agreed to the knowledge to be used in the study. If you take away all the data at that point it would totally undermine the design of the study, number one, and number two, they don't really own the knowledge that comes from their samples. So you can argue for [policy] (d) but as someone who has done research myself I think it would be a very complicated process to set up a study that way (#64 US, medicine, bioethics, university hospital, N).

Respondents against policy (d) argued that the autonomy rights of research participants have limits because in order to practice unbiased science, researchers have to rely on a sort of contract with research participants that permits undisturbed analysis of the data generated during a research project.

> The worst would be [option] (d): (d) says they contributed [their samples] and, by the way, if your researchers relied on this … as long as you have not submitted for publication—tough luck you have to remove, you have to reanalyze. I have heard people suggest this even for clinical trials where it is really ridiculous … it can introduce bias into the study, it can also create a totally unworkable situation … Imagine what the methods would look like. From a science perspective it is bad and …I don't know what the idea would be for autonomy or respect for persons … It is essentially … extending to them a right that is way too much. Basically, they agreed to something, you relied on it, and now they are going to take back on your reliance. That is ridiculous. So I would rule out [option](d) (#10 US, humanities, law, university, R/A).

In general, it is the second part of policy (d) that was most often criticized. The destruction of samples is regarded as less problematic than the destruction of "human genome information" (#62 IS, Europe and No. Am/Europe, medicine, law, university, R).

> [F]or us, the information that concerns the patient is more important. If a patient thinks that all biological material must be removed, it is perfectly fine to remove it (#86 IS, So. Am., medicine, university, A).

2.2 Withdrawal after the death of a research participant

Respondents were divided concerning whether, after a participant's death, his or her legal representative may exercise the right of withdrawal on his or her behalf. Some respondents suggested that it is an issue that must be decided by participants at the time their consent is taken.

More than one third of respondents, both in Europe and North America as well as in other regions, rejected the idea that a legal representative could withdraw consent. Under Federal research regulations in the United States, a dead person does not fall within the definition of a human subject to whom the regulations would apply. Respondents from North America might have been influenced by these legal provisions, although they infrequently referred directly to US law. Many participants from the US rejected the right to withdraw consent of relatives or the legal representative on the grounds "that informed consent is an individual consent; it's not a consent that devolves to someone else" (#95 US, bioethics, philosophy, theology, university, N). Other arguments used against the right of withdrawal being exercised by a deceased participant's legal representatives were that, first, empirical research shows that surrogates are bad predictors of the wishes of patients, and second, that patients who did not withdraw before death apparently wanted to stay in the study.

> First of all, at least in the US, when someone dies they are no longer a research subject and I don't think their legal guardian has any right over that information. They contributed it, they did not withdraw, and according to everything we know the … guardians have no idea what patients [would] want in this situation. It just becomes arbitrary, and it becomes what the legal guardian wants and not what the patient wants unless they can produce some serious, incontrovertible evidence that the patient would have wanted to withdraw. But the patient had already given consent and had not withdrawn before that; I think that's grounds enough (#33 US, bioethics, medicine, government, A).

Other respondents found it even dangerous to give a right to withdrawal to family members due to the possible adverse consequences for research and because this practice would be contrary to the wishes of other research participants who provided their samples to make research advance.

> I am very troubled about the idea of subjects being able to go back to researchers and reach deeply into research projects and pull their stuff out. I feel even more strongly that to expect family members to be able to do that seems absolutely bizarre and socially dangerous and jeopardizes all the efforts that other subjects have put in to participate in the research. So I would be very strongly opposed. Family members might be allowed to do … [option] (b) [anonymization] as form of withdrawal. After a person has died there is no value to include the name, so I don't see a reason not to do it (#59 US, medicine, bioethics, government, R/A/S).

Interviewees against a policy that allows withdrawal of samples after the death of the research participant raised the issue of using samples that have been collected several decades ago.

> If you discover human bones that have been buried 200 years ago, do you have to get the permission from their descendents or not? That's a big question. How many years should have passed after the death of the participant? (#94 IS, Asia, life sciences, government, A).

They pointed out that contacting the relevant representative—a blood relative or a member of the group of which the sample donor used to be part—can be problematic. Yet, they acknowledged that relatives or members of the group to which the research participant belongs may find inappropriate that such samples are used without their consent. Referring to the legal framework in place in his country, one respondent suggested that "an informal rule [stating] that the sample is considered essentially unidentified 50 years after the person is dead ... probably would be best to honour the request from the relatives of the dead person" (#32 IS, Europe, medicine, university, S/A).

On the other hand, a substantial number of interviewees supported the policy granting a legal representative the right of withdrawal on behalf of a deceased research subject. They argued that requests of family members should be taken into consideration because the nature of genetic information in itself implies that individuals other than the research subject have an interest in whether or not samples may be stored and used after the donor's death.

> There is a familial dimension in terms of genetic information, and just the family has to take care of all the heritage of that person (#66 IS, Europe, law, consultancy, A/R).

In the US, respondents argued that possible harm to family members might justify a right to withdrawal. However they were unsure to "how many generations" (#11 US, medicine, bioethics, university, R/A) these considerations apply. Some respondents suggested that a relative's right to withdraw should depend on whether the research results have clinical implications:

> It also depends on the research question. If there are clinical implications to it, then definitely yes [relatives should have a right of withdrawal after the death of the participant]. If there are no clinical implications, I am not sure. We are talking about minimal risk research. In genetics, if there are clinical implications, the clinical implications transfer to family members ... [In the case of] other genes without clinical implications, there is no reason for family members to really say anything about it—it is just biology. If there is a way to make that distinction, which is sometimes very hard, I would say it should depend on the clinical implications (#96 US, genetics, medicine, social sciences, university hospital, R/A/S).

One respondent referred to the intimate meaning genetic information has for some reasonable persons. He thinks that the transfer of the right of withdrawal to family members follows from the humanly justified wish to pass control over important entities to one's heirs after death.

> I'd say it's an inheritable right of withdrawal. So I would say that the family can decide to withdraw ... many people would not care a lot about their genetic information in a deep sense, they would care only about actual material risks like for insurance companies getting the information. But I think there is a subset of people who are themselves reasonable who would attach a lot of meaning to these samples because it's got genetic information about them, and it's a reasonable enough personal view that I want to protect. And once you think of something that's being intimately meaningful to somebody, we tend to think that it is something that your heirs get to continue to have ... control over (#16 US, philosophy, bioethics, university, N).

2.3 Destruction of samples at the end of the research project

The majority of respondents hold the view that the collection described in scenario A should not be destroyed and that a different investigator or institution should become responsible for it. However, most of them think that it is important to inform research participants that their samples will be kept indefinitely. Only a few respondents argued that one should assume in general that, once collected, samples will be available for use in research indefinitely.

Two main reasons were provided in favor of maintaining the collection after the end of the original study: societal benefits and the prevention of scientific frauds. Tissue collections are considered important research tools that may contribute to the greater good by facilitating biomedical research, and destroying them would result in a loss:

> Many people would regard it as being unethical ... to destroy sample banks that could potentially make a contribution to understanding a serious health problem (#82 IS, Europe, natural science, A/R).

Moreover, if samples were destroyed, it would be impossible to verify whether the scientific report was correct. Furthermore, for scientific progress it is important that "in the future possible comparative analysis can be conducted" (#93 IS, Europe/ Middle East, biology, genetics, biobank, S).

Several respondents from the US argued that samples could be used as long as the database is anonymized:

> [T]he whole database should be anonymized. But I don't see anything wrong with that second choice [not closing the biobank. However,] if there is no will or no money for further research, if there is no possible way that things could be controlled or anonymized in the future, then things should be destroyed (#43 US, bioethics, medicine, philosophy, government, O).

The answers of respondents were influenced to a great extent by what they assumed to be the content of the original consent. Respondents in favor of maintaining the collection argued that this could be justified as long as research participants have agreed to future uses because this setting implies that the collection is not destroyed after the original study.

By contrast, most respondents who were in favor of the destruction of the collection at the end of the colon cancer study supposed that the samples were collected for a specific project and that consent was limited to that specific project. They think that a different policy would expose participants to possible harms due to the fact that the "[research] environment can change quite a lot" (#14 IS) or due to "the problem of secondary use" (#46 IS).[12]

2.3 Notable similarities and differences between subgroups of respondents

The issue of participants' withdrawal and the destruction of samples proved to be highly controversial. Interestingly, we did not find any significant differences in the attitudes of respondents from different regions in contrast to the sharp

---

12 #14 IS, Oceania, political science, indigenous IP, university A/R; #46 IS, No. Am., law, social science, NGO, A.

differences concerning consent (see chapters 5 and 6). A similar lack of significant differences was observed with respect to respondents who have used samples and those who have not been directly involved with biobank research. What could be possible explanations for the similar distribution of preferences among the different subgroups? To begin with, sample users do not uniformly favor policies of limited rights of withdrawal, even though such policies might seem to simplify their work. Although anonymization of samples is judged beneficial for research because it allows scientists to continue using the samples, for many this policy is ethically dubious because it seems at odds with the meaning of withdrawal. Moreover, destruction of samples, although complicated, is not perceived as a real threat or burden by many sample users because in their experience, requests for withdrawal are rare. In summary, sample users don't have uniform preferences because they view all alternatives—even policy (a) —as having advantages and disadvantages for researchers. Although policy (a) seems the least cumbersome for researchers, they also fear that it could cause difficulties in recruiting research participants.

The issue of withdrawal seems to be less influenced by cultural differences than the issue of consent because, first, many respondents were unsure about the meaning of withdrawal in traditional research ethics as well as in their own cultural context. Second, many respondents indicated a preferred type of withdrawal, but would not necessarily rule out other types as long as the consent form explained the details of the policy. As explained by several respondents, their preferences on this matter were personal choices rather than something determined by firm ethical presuppositions. Europeans and Americans, as well as respondents from other regions, are torn regarding the significance of autonomy rights with respect to withdrawal, especially since this is a relatively new issue and existing guidelines give only broad and vague indications.

Concerning the destruction of the collection at the end of the colon cancer study, it is difficult to find differences among the subgroups because most respondents' views on this topic depended upon what had been specified in the original consent. Since the respondents adopted varying hypotheses as to the nature of this consent, their answers are difficult to compare. The question of whether recontacting is required or not for future research if the original consent was narrow was not systematically approached in scenario A.8 since it was one of the points raised in scenario D (see Chapter 6).

3. Discussion and conclusions

3.1 Reasons for disagreement and consensus

Disagreement on the meaning of withdrawal For withdrawal, the challenge is not so much to bring ethical requirements for biobanks into line with "classical" health research ethics (as is the case regarding consent). Instead, the challenge is to reach agreement about the meaning of withdrawal itself when it comes to information stored in research databases generally. To what extent have research subjects the right to control the use of their own health information generated during research in which they have been involved? The debate about biobanks has basically added the question about the right to withdraw samples. Although the two interpretations of "withdrawal" are in principle incompatible—that research subjects withdraw by walking away or that they are also entitled to withdraw from a database their samples and information derived from them—almost all respondents acknowledged that the second position is the "stronger" ethical standard and hence would be preferred if it were simple and practical.

Disagreement on empirical questions   Thus, disagreements about empirical questions are in large measure responsible for the variety of attitudes towards withdrawal, and such disagreement might be at least partly resolvable. First of all, no clear pattern exists even among respondents who work with samples. The answers to the empirical question of how frequently withdrawal occurs are shaped by individual experience with particular research projects. According to one respondent, in the Human Genome Diversity Project withdrawal had become a problem. In many other studies only few research subjects asked for withdrawal. The prediction of the frequency of withdrawal is further complicated by the interactions between trust and withdrawal observed by some respondents. If participants are not offered liberal withdrawal options, this failing might decrease trust which in itself will increase requests for withdrawal. Conversely, if withdrawal is assured almost without limits, trust is increased and withdrawal would consequently be rarer.

The burden that withdrawal might impose on researchers varies not only according to the frequency of withdrawal, but also according to the number and type of research collaborations in which the biobank is engaged. Therefore, it is not surprising that interviewees reacted as they did to the question whether withdrawal option (d) is practical or not. Sample users among them—who are used to working with small academic research institutions where close contacts among research groups facilitate collaboration—might find option (d) acceptable whereas it might seem completely infeasible to respondents familiar with research consortia carried out over geographically large and culturally diverse regions, especially if commercial interest and survival of companies are at stake. Empirical data on the feasibility of different types of withdrawal and their impact on different types of research would greatly help to focus the debate on the disagreement on fundamental ethical questions.

Disagreement on the balancing of values   As long as the empirical prediction of the negative consequences of withdrawal on research is uncertain, it seems difficult to reconcile those respondents who give absolute precedence to autonomy rights with those who, fearing grave harm to research, would limit the right to withdraw samples and information at least in all situations where risks for research participants are considered minimal (Eriksson and Helgesson 2005). The first have behind them the entire recent tradition of health research ethics as it is summarized in the Declaration of Helsinki and the WMA Declaration on Ethical Considerations Regarding Health Databases (WMA 2002). Those in the second group who believe that a subject's wish to withdraw from an ongoing study should trigger the anonymization of the subject's samples and personal data are a minority in our study yet are supported by the recent guidelines from the Council of Europe (COE 2006).

Different conceptions of a person's "autonomy rights" after death are the principal reason for the disagreements concerning the right of a legal representative and of family members to withdraw samples after the death of the participant. One extreme is the view that autonomy rights imply ongoing control over samples which means that samples should not be used because the deceased subject is no longer able to exercise control. Others see the autonomy of the research participants as absolute which implies that decisions they made during their lifetime is binding for research after their death. A third intermediate group would attribute control rights over genetic information either to a representative appointed by the research subject before death or to blood relatives who share genetic risks with the research participant. The ethical considerations are further complicated by practical arguments about how to locate family members and legal representatives after the death of sample donors, particularly if the biobank was established many years earlier, and about how to deal with disagreement among various family members about withdrawal. Opinions of interviewees varied widely, last but not least because present guidelines do not provide any help regarding how these questions should be decided.

3.2 Contributions to the current debate

Ulysses contracts are considered unacceptable in classical research ethics: subjects should not be allowed at the beginning of a study to commit themselves to forgoing their right to withdraw themselves in person from an experiment. One of the major questions that remains unresolved is whether Ulysses contracts are acceptable as far as withdrawal of samples and data is concerned since the risks taken by those who sign such an agreement are considered less personally burdensome than if their living bodies were involved. Our study does not provide an answer, but it demonstrates how important it is that these questions are raised and more openly discussed. As one of the respondents noted, most biobanks probably don't dare to say that participants are not allowed to withdraw. However, many studies on biological material stored in biobanks grant the right to withdrawal without explaining what this means for the samples and related information. Guidelines should offer more explicit recommendations about the details of withdrawal policies than they do at present. The silence on this issue in many research protocols and guidelines could mean that their authors don't want to attract attention to this topic for fear of increasing the frequency of withdrawals, similar to the reasoning of one respondent who recommended that biobanks permit withdrawal but avoid encouraging participants to request it.

Although withdrawal, as well as destruction of samples at the end of a study, is a controversial subject, the answers of respondents in our study highlight a number of issues that guidelines should address and that researchers and ethics committee members should take into account when judging whether a withdrawal policy is adequate.

- Whatever withdrawal policy is proposed, it should be explained to research subjects in detail before they give consent to participate in a study.
- The explanations should comprise details on the endpoint that will be used: publication of results is an irreversible endpoint; submission of a manuscript

and the analysis of the final dataset both seem to be reasonable endpoints.

- The explanations should also provide clarification about the right to withdrawal by legal representatives or family members after the death of the research participant.
- The justification of withdrawal policies depends on the type of consent given as well as on the risks related to the research. The broader the consent and the higher the risks, the more extended should be the opportunity to withdraw.
- Access to samples and information needs to be granted for a certain time to control results and to ensure the accuracy of the study. This control is not necessarily a violation of participants' autonomy rights because there will not be any new prospective or retrospective studies on their samples.
- Withdrawal of data during and after analysis ("knowledge" that has been generated) should be distinguished from the withdrawal of samples and information stored in a biobank. As the results of our study show, destruction of the latter, i.e. of samples and data stored in a biobank, is less controversial, on practical and on ethical grounds, than the first, i.e. the destruction of data in the form of results obtained from research involving a biobank.
- The policy chosen for destruction or maintenance of the biobank at the end of a study as well as the length of storage should be explained to research participants and be subject to their consent.

Bibliography

COE (2006), *Recommendation Rec (2006)4 of the Committee of Ministers to Member States on Research on Biological Materials of Human Origin*, Strasbourg, France: Council of Europe.

Eriksson, S. and Helgesson, G. (2005), "Potential harms, anonymization, and the right to withdraw consent to biobank research," *Eur J Hum Genet* 13, 1071-6.

European Society for Human Genetics (2003), "Data storage and DNA banking for biomedical research: technical, social and ethical issues. Recommendations of the European Society for Human Genetics," *European Journal of Human Genetics* 11, Suppl. 2, S8-10.

Harlan, L.M. (2004), "When privacy fails: invoking a property paradigm to mandate the destruction of DNA samples," *Duke Law J* 53, 179-219.

HUGO [Human Genome Organisation] Ethics Committee (1998), *Statement on DNA Sampling: Control and Access* <http://www.hugo-international.org/Statement_on_DNA_Sampling.htm>, accessed July 2007.

Merz, J.F. (1996), "Is Genetics Research 'Minimal Risk?,'" *IRB* 18, 7-8.

NBAC (1999), *Research Involving Human Biological Materials: Ethical Issues and Policy Guidance, Report and Recommendations, vol. 1.* (Rockville: National Bioethics Advisory Commission).

Nuremberg Code Art. 9 <http://en.wikisource.org/wiki/Nuremberg_Code> or <http://www.state.nj.us/health/hrep/documents/nuremburg_code.pdf>.

SAMS (Swiss Academy of Medical Sciences) (2006), *Biobanques: Prélèvement, conservation et utilisation de matériel biologique humain* <http://www.samw.ch/>, accessed July 2007.

UNESCO International Bioethics Committee (2003), *International Declaration on Human Genetic Data* (Paris, France: UNESCO).

World Health Organization (European Partnership on Patients' Rights and Citizens' Empowerment) (2003), *Genetic Databases, Assessing the Benefits and the Impact on Human Rights and Patient Rights*, Geneva.

World Medical Association (2000), *Declaration of Helsinki* <http://www.wma.net/e/policy/b3.htm>, accessed July 2007.

World Medical Association (2002), *The World Medical Association Declaration on Ethical Considerations Regarding Health Databases* (Ferney-Voltaire, France: World Medical Association).

Chapter 9

Anonymization and Coding

Bernice Elger

1. Introduction

In this chapter we will provide a summary of the attitudes of participants in our study towards the ethical questions raised by the anonymization and coding of samples and data stored in genetic databases.[1] Participants' responses will be presented in the context of the ethical debate in the literature and the positions taken by important international and some national guidelines on this issue.

1.1 Anonymization and coding in the context of genetic databases

Sample donors and the public are increasingly worried about the sensitivity of genotype information and the management of samples and data (Hoeyer et al. 2004). They ask for stricter control because of the fear that their privacy is threatened if data protection is insufficient. The public is concerned about unauthorized electronic access to data files and the use of stored information for illicit purposes (Clayton 2005) and is apprehensive of the risk that personal genetic information is transmitted to third parties such as insurers and employers. It is well known that both insurers and employers could under certain conditions put pressure on individuals to release information contained in their medical records or information generated by their participation in medical research.

In this context it is not astonishing that the question as to which form of anonymization or coding should be chosen for the storage and use of samples and data plays a central role in the ethical and legal debate.

1.2 The ethical consequences of anonymization and the terminology problem

The debate about anonymization and coding is framed by the Declaration of Helsinki, which states that "Medical research involving human subjects includes research on identifiable human material or identifiable data" (WMA, 2000). It could be concluded that any research using non-identifiable samples does not need to meet the ethical and legal requirements of medical research such as informed consent and IRB approval.

---

1 In this text, the terms "genetic database" and "biobank" are used synonymously, see the explanations in Chapter 1.

However, an analysis of the literature for a definition of "identifiable" reveals that the meaning of "identifiable" is anything but clear. A multitude of different terms exist (see Elger and Caplan 2006, Knoppers and Saginur 2005), and almost every guideline uses a distinct terminology. Examples of terms used frequently are: anonymous, anonymized, completely anonymized, unlinked or linked anonymized, reversibly or irreversibly anonymized, linked or unlinked to an identifiable person, irretrievably unlinked to an identifiable person, de-identified, permanently or not permanently de-linked, not traceable, identifiably linked, pseudoanonymized, encoded, encrypted, confidential, identified, nominative, directly identified, fully identifiable, personal data. Communication barriers are the unavoidable consequence where the same term is used with a different meaning in different guidelines as is presently the case (Elger and Caplan 2006).

In this chapter, the following terminology is used (COE, 2006a). The term "anonymized" means that biological material is stored alongside associated information, such as the type of tumour, medical treatment, donor's age and so forth, but all information that would allow identification of the research participant or patient is stripped, either irreversibly (unlinked anonymized samples) or reversibly (linked anonymized or coded samples). Finally, samples are considered to be identified if the information that allows identification—name, address and so on—is associated directly with the tissue, such as when the patient's name tag is attached to the sample.

1.3 Brief outline of the current debate on the concept of identifiable data and samples

Discrepancies are not limited to different terminologies, but even more seriously, involve the whole regulatory framework. Originally, identified, coded and reversibly anonymized samples and data were considered as identifiable in most recommendations (COE 2006a, NBAC 1999, WMA 2002, WHO 2003, HUGO 1998). In all these cases a link exists. Only if this link is irreversibly destroyed are samples and data considered unidentifiable and, thus, research using such samples was not considered human subject research in accordance with the Declaration of Helsinki.

The distinction between identifiable and non-identifiable data and samples has been challenged recently in two different ways. First, it has been stated that non-identifiable samples do not exist because it is impossible to achieve complete anonymity for individual genetic data since DNA fingerprinting and cross-linking of different databases might permit identification of sample donors (Lin 2004). Moving exactly in the opposite direction, the US Office for Human Research Protection (OHRP) broadened recently the definition of "non-identifiable" (OHRP, 2004) in the following way:

> OHRP considers private information or specimens not to be individually identifiable when they cannot be linked to specific individuals by the investigator(s) either directly or indirectly through coding systems (OHRP, 2004).

This is the case if "the investigators and the holder of the key enter into an agreement prohibiting the release of the key to the investigators under any circumstances, until the individuals are deceased (note that the [Department of Health and Human Services] regulations do not require the IRB to review and approve this agreement)." It is also the case if "there are IRB-approved written policies and operating procedures for a repository or data management center that prohibit the release of the key to the investigators under any circumstances, until the individuals are deceased" or if "there are other legal requirements prohibiting the release of the key to the investigators, until the individuals are deceased." OHRP also specifies that "[t]his guidance applies to existing private information and specimens, as well as to private information and specimens to be collected in the future for purposes other than the currently proposed research." The advantage of enlarging the definition of non-identifiable is obvious: researchers are provided with a simple means to escape strict regulations by entering agreements that prohibit them from access to the code, without having to destroy the link. Through these simple arrangements, any type of future research in the US is authorized without the need for consent or IRB approval (see Elger and Caplan 2006).

The resulting differences between the US regulations and guidelines from other parts of the world concerning the concept of "identifiable" samples and information could hamper international research collaborations (see also chapter 5.).

1.4 Types of coding: published recommendations

Although data protection is strongly influenced by the scientific advances in bioinformatics and in spite of the new technology constantly created with the intention to resolve privacy issues (Quantin et al. 1998; Knoppers and Chadwick 2005), overall, the literature on the ethical implications of technical details of coding is still scarce.

According to Auray-Blais and Patenaude (2006), some research ethics committees (REC) have refused to accept research projects involving biological samples from a biobank because samples were coded and not irreversibly anonymized. The RECs were concerned that coding implied a risk of discrimination and stigmatization through uncontrolled use of the samples or any related genetic data.

International guidelines and the recommendation from the NBAC[2] Researchers looking for guidance find only limited help in international guidelines that address the issues related to biobanks (see also chapter 3). All of these guidelines stress without exception the importance of protecting the privacy of the research subjects and the confidentiality of samples and data (WMA 2002, WHO 2003, HUGO 2002). However, few details are given on the coding procedures. The UNESCO guidelines

---

2 As already mentioned in chapter 3, we will include here the guidelines from the US National Bioethics Advisory Committee (NBAC) because they are relevant for the large number of biobanks that are in operation in the United States as well as for international projects with US participation. They also form the normative framework for half of the experts interviewed in our study.

include the general recommendations that "the necessary precautions should be taken to ensure the security of the data or biological samples" and that "data and biological samples linked to an identifiable person should not be disclosed or made accessible to third parties, in particular, employers [and] insurance companies" (UNESCO 2003 Art 14). The Council of Europe requires rules in place to ensure confidentiality without specifying these rules:

> Clear conditions governing access to, and use of, the samples should be established ... Quality assurance measures should be in place, including conditions to ensure security and confidentiality during storage and handling of the biological materials (COE 2006a, Art 4.).

The reason for the lack of precisions in the COE guidelines is given in the associated preliminary draft explanatory memorandum. Biobanking is a fast moving field. Therefore, the guidelines should be adapted to changing circumstances.

> This allows in due course for in depth reflection on the appropriateness of more detailed provisions in the recommendation (such as on arrangements for coding or anonymization, or on commercial use of biological materials of human origin) (Draft COE 2006b, Art. 27).

The NBAC requires the description of the protective measures and leaves the judgement whether the protection of confidentiality is adequate to the IRB. Researchers are under the obligation to provide "a description of procedures used to minimize risk to subjects" and a "full description of the mechanisms that will be used to maximize the protection against inadvertent release of confidential information" (NBAC 1999, recommendation 5).

The European Society for Human Genetics insists that the goals of coding are not only to protect confidentiality, but also to permit standardization and sharing of information. Security mechanisms to ensure the confidentiality and long-term conservation of genetic information should include "standardization of coding, sample tracking, computerization and encryption. The standards adopted should permit sample information sharing for research purpose with minimum risk. Discussions should be encouraged among consortia of bankers to issue standardized bank protocols" (European Society for Human Genetics 2003, p. S9).

The WHO recommendations are noteworthy first because they introduce the concept of "proportional or reasonable anonymity" and, second and more importantly, they contain the requirement of an independent instance in control of the coding. Coded information is considered reasonably anonymous "when access to the link is restricted appropriately."

> In keeping with international standards of anonymization, as laid down, inter alia, in the *EC Directive* on the Protection of Individuals with Regard to the Processing of Personal Data and on the Free Movement of Such Data, it is acceptable that proportional anonymity be used to secure genetic samples or genetic data, depending on how this is done, who is to have access and the uses which have been consented to (WHO 2003, recommendation 7).

To ensure that anonymization procedures are adequate and protect individual interests, these guidelines contain the recommendation "that any anonymization process be overseen by an independent body" which functions "as an intermediary between the creators and the users of the database" and has the obligation to "scrutinize and ensure the legitimacy of requests to the database" and to "keep anonymization processes under review" (WHO 2003).

A comparison of the international guidelines reveals agreement that *coded* data should be used "in preference to readily identifiable data" (WMA 2002, Art 24.). However, some controversy exists concerning irreversible anonymization. Whereas the UNESCO and WMA guidelines recommend the use of irreversible anonymization "normally" (UNESCO 2003) or "wherever possible" (WMA 2002), HUGO advises that "careful consideration should be given before proceeding to strip samples of identifiers since other unknown, future uses may thereby be precluded as well as may the validation of results" (HUGO 1998) and the NBAC requires "special precautions" if samples and data are irreversibly anonymized, including IRB approval to ensure that "the unlinking of the samples will not unnecessarily reduce the value of the research" (NBAC 1999, recommendation 3). Among the reasons why NBAC insists on controlling irreversible anonymization is the following:

> [I]nvestigators sometimes may be tempted to choose unidentified or unlinked samples in order to avoid the more stringent standards and procedures required for coded or identified samples—for example, the requirement for measures to protect privacy and confidentiality (NBAC 1999, p. 43).

Finally, the international guidelines vary with respect to the importance they attribute to the respect for the choices of research participants concerning anonymization. The HUGO guidelines contain the most explicit recommendations on this issue. HUGO insists that "the choices and privacy of individuals, families and communities should be respected" and explains that "choices may be with regard to: donation, storage and uses of samples and the information derived therefrom." However, more detailed recommendations are lacking. Although advising that "[m]echanisms should be established to ensure respect for such choices" (HUGO 2002, Art. 4.), HUGO does not explain what these mechanisms are. Moreover, it is not clear to what extent participants should, according to these guidelines, be offered the possibility to choose actively between various types of coding. The only condition specified by HUGO in this respect is that "[p]articipants should be informed of the degree of identifiability of their data (e.g. coded, anonymized, aggregate, etc.) and of the security mechanisms in place to ensure confidentiality" (HUGO 2002, Art. 4.).

Other recommendations Several guidelines from national bodies or professional organizations contain more detailed recommendations including an independent key-holder and double coding. The ethics committee of the ASHG published the recommendation that researchers "consider a way of coding samples by a third, independent party, who would keep the codes inaccessible unless there are specific circumstances in which the code needs to be broken" (Godard et al. 2003). The Ethics Committee of the Swedish Medical Research Council recommended that the codes permitting identification of an individual should be kept by a trustworthy public institution such as a university or medical authority (Abbott 1999). In 2004, the European Agency for the Evaluation of Medicinal Products (2004) maintained not only the idea of an independent third party functioning as key-holder for the codes, but added the category of double coding. Samples are coded by a first series code, and a second code is assigned to link the samples to the results, i.e. the information obtained from the samples. In order to identify the person who provided the sample, it is necessary to combine the two code keys. A similar form of double coding has been recommended recently by a Canadian group (Auray-Blais and Patenaude 2006).

1.5 Is double coding, single coding or irreversible anonymization adequate for a typical long-term research biobank (scenario A.1)?

The issue of different types of coding versus irreversible anonymization was illustrated in scenario A.1 (see chapter 4). The vignette describes an international biobank established to enable research on the association of certain genes with an elevated risk for developing colorectal polyps (CP-biobank). Apart from samples, the biobank stores information about health and lifestyle of patients and their family members that will be periodically updated. Many important contemporary biobanks use a form of reversible anonymization because this is a way to ensure protection while keeping a link to be able to update information and to recontact participants if information is discovered during the research project that could be of value for the participants. The policy that was proposed in scenario A.1 was modeled according to existing banks in Canada and Switzerland (Swede et al. 2007, Grotzer et al. 2003). The protection proposed in scenario A.1 is a form of double coding. The physicians who collect the samples code the samples and associated information by assigning a number to each participant (1st-Series Code). Researchers and the biobank itself have access only to samples and data labeled by a second code. The transformation from the first to the second code is under the control of an independent authority (described as independent scientist). Once each sample and the accompanying data have been received by the biobank, this scientist not otherwise involved in the operation of the repository removes the existing code and assigns a new number to each sample and associated information (2nd-Series Code). The scientist keeps a list of the 1st-series codes and the corresponding 2nd-series codes in a secure location outside the repository.

We were interested whether the experts we interviewed think that the double coding adequately protects the interests of the participants, whether such a coding system is perceived as being necessary or whether the coding process could be left entirely in the hands of the repository, or, finally, whether any form of coding should rather be abandoned in favor of irreversible anonymization of the samples and associated information.

2. Results

2.1 Irreversible anonymization

We report on the results concerning irreversible anonymization first, because it is one of the few issues in our entire study where almost all respondents agree. With a few exceptions, respondents from the international and from the US samples do not favor irreversible anonymization of samples and data in scenario A.1. Whereas several respondents admit that irreversible anonymization can be justified in other types of studies, they believe that in the case of the CP-biobank irreversible anonymization would preclude the realization of the study purpose which includes adding follow up data related to the health of patients and their family members. Another reason mentioned was the impossibility to permit participants' withdrawal from the study if the link is destroyed. Several experts from the US sample as well as most of the non experts insisted on the importance of feedback of research results that would become impossible in the case of irreversible anonymization. Some interviewees from the US insisted on the right of study participants to decide whether they preferred irreversible anonymization or coding. The few respondents in favor of irreversible anonymization said that this was the only way to protect confidentiality efficiently and that according to their knowledge most study participants felt more comfortable with irreversible anonymization because data security was maximized.

2.2 The type of coding

The answers of respondents showed controversial attitudes towards the following questions. Should double coding be required or is single coding acceptable? Should an independent party be the key-holder or would it be acceptable that the repository carries out the coding and keeps the code keys? Is the key-holder role of physicians concerning the first series code adequate or rather problematic?

Almost all respondents from the international sample and somewhat less (two thirds) of the respondents from the US sample agreed that the double coding does adequately protect the interests of study participants in scenario A1. Interestingly, three quarters of the international respondents, but only less than one third of the US experts answered that this form of double coding system is necessary. The few interviewees from the IS sample who said that double coding is not necessary were almost all from Europe.

Only one third of respondents from both samples think that it is necessary to involve an independent party to keep the code keys such as the independent scientist described in scenario A.1. However, although approximately two thirds of both the international and US experts said that it is acceptable to leave the coding entirely in the hands of the repository, many of these would, if given the choice, prefer the option of involving a key-holder who is independent from the repository. A sizeable number of respondents from the US sample, predominantly experts directly involved in the use of samples, expressed the opposite view, i.e. that it is preferable to have the coding done by the repository without involvement of an independent party or physicians.

2.2.1 Arguments in favor of double coding and an independent key-holder Respect for privacy and confidentiality, both sub-principles of the respect for autonomy, were the most often cited as arguments by interviewees who think that double coding is necessary.

Some respondents think that the respect for research participants' autonomous choices is the principal argument in favor of double coding, independently from the empirical protection provided by double coding. According to these respondents, double coding honours best the wishes of sample donors to have their privacy protected in a trustworthy way. The assumption is that wishes motivated by fears should be respected even if the fears are exaggerated and based on an overestimation of existing risks.

[P]eople in Africa ... are very hesitant to give their personal information to other people (#22 IS, Africa, science, self-employed, A).

I think there is a bright fear among [my indigenous group] of DNA and for [this indigenous group] there is something special about their DNA and a lot of them do not want their DNA to be taken and linked to others, most of them because we are not a homogenous group (#37 IS, Oceania, social sciences, bioethics, university, A).

Some disagreement exists about the empirical question as to what the wishes of research participants are. A respondent from Oceania said he[3] had observed "that more and more patients are not as worried about that as they used to be if the research is going to be used and reported back to them" (#14 IS, Oceania, political science, indigenous group, university A/R). This respondent is in favor of double coding, but thinks that the coding can be left entirely to the repository because he is convinced that for research participants, privacy is not always the main aim but is outweighed by the interest to receive research results.

Most interviewees referred to consequentialist arguments in order to balance ethical principles or made reference to the principles of beneficence and non-maleficence which both imply benefit-risk evaluations.

To begin with, the most frequent reason mentioned in favor of the double coding system in scenario A.1 is *its expected benefit*: Double coding "protects confidentiality ... better than single coding" (#50 US, catholic theology, genetics, philosophy, bioethics, university, A). It is said to "magnify" security (#52 US, social science, bioethics, university, R/A). Respondents in favor of the double coding system are convinced that this was the best way to maximize confidentiality without sacrificing the objectives of updating samples and data and feedback to participants.

So I think that the benefits of the double coding substantially outweigh the complications (#92 US, law, medicine, genetics, university, R/A).

Second, the *avoidance of risks* was a central argument in favor of double coding. Respondents' risk evaluations are influenced by different cultural and socio-economic contexts. The reasons why so many respondents from the IS think that double coding is necessary seem to be related to particular concerns about the abuse of information. The risks were described in two forms, as subjective fears which might not reflect

---

3 In all following chapters, to simplify, all respondents, independently of their gender, are referred to by male pronouns.

reality (see above the quotation from respondent #37) and as possible objective facts as shown in the following example:

> In a situation where the disclosure of private information would have an impact on for instance employment or insurance, I think that the more protection is there the better. I think that's the criterion (#56 IS, Europe/Europe and Africa, medicine, life sciences, government, hospital, S/A/R).

Several respondents from the US sample indicated that double coding with an independent key-holder is necessary because they consider that conflicts of interest can lead to abuse:

> If you had only one person it could not be sufficient in an anomalous situation where some person was pushing very hard to get information or just in terms of social network. Very often social networks overlap in such a way that what looks like clear separation of interests turns out not to be and if you add another layer in there: you just magnify the [security] (#52 US, social science, bioethics, university, R/A).

> I think it [the double coding system] is necessary. It seems reasonable to me, there might be other alternatives … What I like about this is the independent key holder … Independence is important because it prevents conflicts of interest (#20 US, law, philosophy, university, R/A).

Respondents from the international sample are convinced that in particular the international character of the repository creates conflicts of interest for researchers, for example to publish new results about genetic risks of populations from particular countries which might stigmatize these populations. These conflicts of interest might put participants even more at risk if researchers could find out the identity of those groups of sample donors who have gene sequences that are particularly interesting for further research because these sample donors might then be recontacted and are vulnerable to exploitation.

> [W]hen you are looking at large groups of samples and international researchers, then you might have some conflict of interests, for such studies when you have association of genes and you don't have a straightforward genetic disorder but you have a gene and a mutation and a risk, you have something more because it is a multifactorial disorder—so the prediction of risk might not be that direct—and therefore it would be better to protect the [participants] because you have different countries and different researchers involved (#19 IS, Europe/Middle East, genetics, university, S).

In line with this, a respondent from Africa specified that double coding is necessary only if the repository "is sharing the resources" with researchers from other countries (#81 IS, No. Am./Africa, medicine, bioethics, university, S).

Double coding might be chosen for the benefit of the repository. As one respondent from Europe (#62) puts it, "it is not a matter of right or wrong, but what is going to be politically better" for researchers and the repository because it increases the chances to find individuals willing to provide samples:

And if these people will feel more at ease and will be more willing to participate in research, one might want to sponsor that additional cost and have that [double coding] happen … for example in the United States, it is very hard to recruit volunteers from Ashkenazi Jews to participate in genetic research, and the main reason is that they are concerned with confidentiality and privacy and getting insurance and employment and everything. So if this would ease anxiety, then I would go for it. On a personal basis, I don't think it is really necessary but if that's going to help the researchers and the repository to work, then I would opt for that (#62 IS, Europe and No. Am/Europe, medicine, law, university, R).

Finally, several interviewees referred to what could be called the "principle of caution":

I certainly like to err on the side of more protection rather than less (#70 US, genetics, social science, bioethics, university, R/A/S). Erring on the side of caution it is certainly good to use this kind of double blinding as the gold standard (#98 US, philosophy, medicine, genetics, medicine, university, R/A/S).

2.2.2 Arguments against double coding and an independent key-holder and in favor of simpler coding procedures

Respondents against the double coding described in scenario A.1 came from two opposite groups: those who preferred simple coding whose arguments will be summarized here and those who indicated that different, more complex procedures are necessary to assure adequate security of the coding. The arguments of the latter will be described below in 2.2.3.

Principle-based arguments   A comparison of the answers of those who thought that the double coding system with an independent key-holder is necessary with those who find less complex simple coding systems acceptable or even preferable shows that the groups hold different opinions regarding research participants' wishes. Several respondents are against double coding because they believe that participants would prefer single coding.

I would [be] in favor of simple coding. I can't understand why people would be so concerned about wanting to be so anonymous [referring to double coding] (#63 US, bioethics, medicine, catholic theology, university, A).

In addition to the respect for participants' choices, both groups referred to confidentiality as an important value, but differ with respect to the way in which confidentiality and the respect for privacy are balanced against other values perceived as important. Among such other values judged more important than confidentiality by respondents against double coding was the efficiency as well as the quality of research. A sizeable number of respondents using or storing samples are worried about the possibility that a second code is "burdensome" for the repository (#53 IS, Europe, medicine/public health/epidemiology, international organization, R/U/A).

A first series code is sufficient … The potential for … more problems in coding for me is also unethical; it is unethical to do bad quality research and anything that increases the potential that your research will be of poor quality is not ethical conduct of research (#96 US, genetics, medicine, social sciences, university hospital, R/A/S).

"Data availability" was another value to which interviewees made reference. The coding system "does not explicitly include the participants in access to relevant data." For the protection of confidentiality "it looks OK, but confidentiality is not the major principle" (#51 US, Europe, bioethics, philosophy, theology, O).

Risk-benefit evaluations in line with the principles of beneficence and non-maleficence, were part of an important number of arguments used against double coding. Many respondents from the US sample are not in favor of double coding because they trust their research system to provide sufficient protection through simple coding. Risks are perceived as rare and as being related to particular circumstances, for example risks related to the feedback of study results in particular circumstances. As the quotations show, for these respondents, simple coding is sufficient in scenario A because the feedback of results in the study on colorectal polyps is not of noteworthy ambiguity, the disease studied is not particularly sensitive, and the samples are judged as non controversial:

> I think it is useful to have a system like that [in scenario A.1] if there is a lot of ambiguity whether you would give results back to individuals. If you think … some people might be tempted to and you want to ensure against that [this would be an argument for] a system like this in place. But I generally think that people are [trustworthy] and having a coding system and a protocol in place to ensure confidentiality seems to work adequately without having the double coding in place. In most cases simple coding would be fine (#26 US, medicine, social sciences, government, R/A).

> Maybe if there is a case where you have special worries about the samples, maybe not colorectal [polyps] but something else. I'd say it [double coding] might be a good idea in that case but I'd say as a general rule it is not necessary to have the second layer (#27 US, philosophy, bioethics, government, R/A).

> I am of the view that you don't always need a double blinded double coded; anonymity of that level is not always required because I am not sure that all samples are all that controversial or all that troubling in terms of the consequences or the factors that might develop (#98 US, philosophy, medicine, genetics, medicine, university, R/A/S).

According to a respondent from Europe (#48 IS, No. Am./Europe, philosophy, university, R/A) the type of coding system should depend on the type of repository. Only in the case of a commercial repository is the independent scientist necessary. If the repository is within the public health system as in scenario A.1, an independent key-holder is not required.

Respondents against double coding differ from respondents in favor of it concerning the situations suspected to create conflicts of interest. For example, one US respondent against double coding thinks that it would be particularly risky if researchers generated the samples, which is not the case in scenario A:

> I don't think so [that double coding is necessary] though it is used quite widely … I think double coding is especially important when the person generating the sample is also going

to be the researcher; then it would have to be double coded I think (#41 US, bioethics, law, university hospital, R/A).

A respondent from the international sample, on the contrary, described the risks in the opposite way: double coding might interfere with the relationship between participants and researchers which is perceived to be of major importance. However, these risks are not judged of sufficient importance by this respondent, to outweigh the necessity of double coding.

> I am worried that [complete] confidentiality [resulting from double coding] means that there is no on-going relationship with the donor and the researchers. In my view, it is very important that a relationship needs to be continuous ... the donor must be contacted. This system is ok because it is a form of protection but there must be a continuing relationship with the donors (#77 IS, Oceania, social sciences, NGO, A).

Another empirical issue related to risk evaluation had a significant influence on respondents' attitudes. Several respondents, from the US sample involved in storing or using samples were against double coding because, according to their experience, the risk of errors is significantly increased if two codes exist for the same sample. Some respondents had themselves witnessed errors in research results because samples were miscoded or data were lost and stressed that these errors are not in the interests of participants who would prefer to reduce errors concerning their samples.

> I think multiplying the codes is dangerous for the research because you can have mistakes along the way. I think that it's important to separate immediately the coding of the samples from all the nominal information. That's the most important part. And after that, if you multiple the codes, I am not sure it is very useful (#57 IS, No. Am./Europe, medicine, university, S).

In comparison, interviewees in favor of double coding did not share this risk evaluation:

> No, I don't [think that double coding causes errors]. It could, but you could get errors even with a single code. I think it is an empirical question what the error rate is and I would imagine that the error rate is not only related to the approach to coding but the sophistication of the employees and the system you are using if they are using bar codes for all this information and everything was being done in a computerized way then they can probably have less errors with a double code than others would have with a single code done by hand (#59 US, medicine, bioethics, government, R/A/S).

Another respondent who uses samples, but is in favor of double coding, thinks that the risk for errors is acceptable for two codes, but becomes non-acceptable only in the case of three codes (#78 IS, Asia, life sciences, government, S/A).

One respondent from South America fears that in the case of double coding one of the codes might be lost and with it the information (#86 IS, So. Am., medicine, university, A). However, he considers double coding necessary in order to ensure confidentiality and is willing to accept the risk of errors.

A significant part of the disagreement between respondents in favor and against double coding seems to be due to different evaluations of the empirical question how much security is provided by various coding systems. Predominantly respondents from US government-funded research institutions, but also respondents from the international sample who use or store samples, think that double coding is "overkill" because it does not guarantee better protection. According to these respondents, the "complications" of double coding outweigh the insignificant increase of protection.

> For me the second code it is a little extra complication without adding significant extra protection (#27 US, philosophy, bioethics, government, R/A).

> It goes back to my uncertainty whether you really need double coding or not and I can't really answer that. I guess I would want to examine it more whether really it is an extra protection or just the appearance of a greater protection, if it is just the appearance and it makes things more inconvenient I would think as long there is minimal chance that privacy is breached, then I would think you don't really need that extra step (#45 US, bioethics, philosophy, government, O).

The same argument that has been used in favor of double coding, i.e. the respect for participants' choices, was also put forward against double coding. A few respondents, all from the US, find obligatory double coding inappropriate. These respondents indicated instead that it is "up to the source of the samples to decide" (#38 US, bioethics, philosophy, theology, medicine, R/A).

> [The double coding] does not adequately protect the interests of the participants, they would need to do various things to agree, some of them may not want the coding ... I don't care whether it is coded or not, it is up to the source of the samples to decide whether he is worried about coded or not. [... Double coding is necessary] for those who want it, yes. I have never met any anybody who felt strongly about having two levels of coding but I suppose if people were presented with the idea there would be some sufficiently nervous about coding that they would be concerned (#38 US).

That respect for research participants' autonomy was also the reason cited by respondents from the US sample to explain why the independent key-holder is not necessary. Although it is a less desirable coding option, leaving the coding entirely to the repository is ethical as long as research participants have been informed:

> [I]f the repository keeps the key then there are other questions that are raised. The issue is, it is a separate question: is it ethical or is it desirable and I think it's better if the key is kept by an independent person because they would vouch for the legitimate use of the key and you would have to apply to the key-holder to get permission, but having the repository hold the key I don't think is necessarily an ethical violation so long as the informed consent to the sample donors indicated that (#41 US, bioethics, law, university hospital, R/A).

Practical arguments Practical considerations mentioned as arguments against double coding are related to costs. One respondent working in Africa thought that the double coding system is adequate but added that Africans lack the resources lack to pay for double coding. Therefore, the biobank for which he works assigns the coding without the involvement of an independent key-holder (#56 IS, Europe/ Europe and Africa, medicine, life sciences, government, hospital, S/A/R). Such budget limitations were of practical importance also in Europe:

> Yes I think it [double coding] is much preferable to ... just one code ... I would not say it is necessary, if you don't have the budget to do it; I am not saying without it you would not be allowed to have a databank, but if you can plan it that way I think it is preferable (#39 US, Europe, philosophy, bioethics, government/university, R/A).

A respondent from South America would leave the coding entirely to the repository because the implementation of an independent scientist is a practical problem:

> It is difficult to find a completely independent scientist to do it. It is a practical reason (#49 IS, So. Am., biology, bioethics, university, A/R).

A few interviewees evoked practical considerations related to the role of the physicians as arguments against the coding system described in scenario A.1. Respondents from the US sample are worried that the role of physicians in the coding process is not adequate because the fact that physicians assign the first series code constitutes a security risk.

> I wonder whether that [the fact that physicians are responsible for the first code] is asking a lot of physicians and if they do it whether physicians have the infrastructure to track; that that was another practical consideration (#26 US, medicine, social sciences, government, R/A).

These respondents from the US were concerned that the link to patients might get lost because physicians are not set up to keep track of the coding. They might retire, or lose contact with their patients. In addition, criticism was voiced that a system having individual physicians attribute codes implies a risk to confidentiality because physicians could be identified by their unique ways of coding.

> You would lose some degree of anonymity if you could identify who the coding physician is maybe by the number of digits they used (#41 US, bioethics, law, university hospital, R/A).

The attitudes towards the role of physicians reflect cultural differences. A respondent from Asia defended an opinion almost diametrically opposed to the statements from US respondents. He criticized that the role of the physicians in scenario A.1 is not extensive enough. They should be responsible for the entire coding, because only then would the coding process be in line with the practical functioning of the health system of the country.

> In the context [of my Asian country], the doctor is a very important person with respect to the patient. The ethical value is very different from the way the Western world looks at it. The doctor is the closest individual beyond the husband and wife. That's the way in [my Asian country] we look at the doctor. The doctor is the depository of all my suffering so we do not consider that information is confidential to the patient or to the doctor, it is responsibility of the doctor to see that the coding is done so that the researcher does not

understand who the patient is. Unlike the Western [world], where the patient is a purely anonymous entity, just an individual who has a social security number, here it is not— the doctor knows the family, knows the patient, the doctor is part of the social system. And that's the way the system works [in my Asian country] (#78 IS, Asia, life sciences, government, S/A).

2.2.3 Arguments against the double coding described in scenario A.1 and in favor of more complex coding procedures

Insufficient protection   Several respondents criticized the coding system described in scenario A.1 because it might not sufficiently protect the privacy of research participants. One respondent from South America was worried that the double coding is not secure because the individuals involved in the coding might not all respect the rules. The respondent added that the double coding is only adequate "if all people do what they are required to do, for instance by not passing on material without permission" (#06 IS, So. Am, philosophy, bioethics, university, A). Similarly, a respondent from Asia (#13a IS, No. Am./Asia, philosophy, university, A/R) was concerned about the integrity of the independent key-holder and asked who will supervise this scientist.

It is noteworthy that several respondents who had particular experience with biobanks or social science research were concerned about datasets that permitted identification of individuals with special characteristics despite the coding and the absence of any name or directly identifying item.

> The second thing we insist on—and I know there is some controversy about it—is that we insist on re-coding the samples when we send them out. And that is because we don't want it to be obvious to other persons doing different studies that, if they compare data, they see that they received samples from the same person. If we use the internal lab code when we send them out to two different labs, the data can be combined without us knowing anything and there is an issue of the so called 'backward identification'. In our country, there is a formal guideline that if there are less than eight persons that have that sort of characteristics, then the sample is identified. For example, if the sample is coded but it is known that the patient had cancer, and the diagnosis date, and that the person was from [my country], and that it is likely that there are less than eight persons, then it is identified by definition. If we send it out with the internal biobank code, with some accessory information that leads to different people, we cannot control people beyond our control who try to combine and perhaps arrive to the identities by for example reconstructing that in one study the information is that the patient has liver cancer and in the other study there may be information that [the patient] was female. And if this information is added and combined, it can result in backward identification. So this is slightly controversial since some people think that different people, working with samples … from a biobank, should be able to combine the data in order to see if study A and study B are the same … .but we insist on re-coding them before they are sent out (#32 IS, Europe, medicine, university, S/A).

Procedures for re-identification   Several interviewees from the international as well as from the US sample expressed the opinion that the crucial issue is not whether single or double coding is used and whether an independent scientist controls the coding or not. Instead, the acceptability of the coding system depends on procedures put into place to regulate re-identification.

> In a carefully run system, a single code would be just fine, but it depends on who does the coding, what are the criteria for using the coding to re-identify. I think people should decide when setting up these kinds of system under what circumstances would someone do the re-identification, what would be a good cause for that. Second, what would be the procedure for that ... and what are the constraints for that—have they signed a legal agreement that says 'in doing that, I would not abuse, I would not do things that I am not supposed to' (#68 IS, No. Am./Europe, nationality: No.Am., life sciences, consultancy R/A).

2.3 Notable similarities and differences between subgroups of respondents

As explained before (see chapter 5), in this part we will compare two types of subgroups, first the two thirds of US-European[4] versus all other international respondents and second, the 31 respondents who used or stored samples versus all other respondents with experience in making recommendations and analyzing ethical and legal issues without "practical" involvement with samples.

First of all, both the US-European respondents and the other international respondents agreed on a number of issues, such as the inappropriateness of irreversible anonymization in scenario A.1 and the importance of preventing adverse consequences for study participants by the way of protecting the confidentiality of the assembled data. The majority of both groups said that the coding policy in scenario A.1 adequately protects research participants' interests. Both groups identified conflicts of interest as a threat to the protection of research participants' confidentiality. Nevertheless, a majority of both the US-European respondents and the other international respondents think that it is acceptable to leave the coding entirely to the repository, although many consider the involvement of an independent coding instance preferable if the resources are available to implement such an instance.

Differences between both groups exist concerning the question whether double coding is necessary and for what reasons. US-European respondents are less in favor of double coding than the other international respondents. The attitudes of the latter seem to be based on the context-related empirical evaluation that simple coding is less secure than double coding. A safer system was judged necessary because fears of abuse among populations in Africa and Asia are high and trust in the research system is lower than it seems to be in the US-European countries. An international biobank storing and using samples from three countries seems to be perceived as threatening in cultural contexts where trust is built through direct relationship with physicians and researchers.

Interviewees with personal experience in using or storing samples seem to be somewhat more critical towards the coding system proposed in scenario A.1 than respondents who never worked with samples. The former predominantly commented on the practical details of various coding procedures. They pointed out

---

4 Since almost all North American respondents in our study are from the US, we utilize the abbreviation "US-European" for the country group including the US, Canada and countries from Western Europe.

the risk of errors and were particularly concerned about the financial and practical burden of complicated coding procedures. In their arguments, these interviewees referred typically to the limited evidence of their personal experience: "we don't do this, because…" or "in our experience it is better to…". On the contrary, as one would expect, respondents with a background in ethics or law were in general more elaborate concerning the theoretical analysis of balancing rights and values and did only sometimes discuss the empirical assumptions about coding details that influenced their positions.

3. Discussion and conclusions

What are the common grounds and irreconcilable differences between the arguments in favor and against different types of coding?

3.1. Disagreement on empirical questions and balancing of values

According to our analysis, a substantial part of the differences of respondents' attitudes is due to disagreement on empirical evaluations. The questions which benefits and risks exist and how high both are require empirical answers. Both risk and benefits are often difficult to predict and vary in different socio-cultural contexts. Therefore room is left for some variation of appreciation. The answers of several interviewees show how these empirical uncertainties influence attitudes. The risks of abuse are perceived to be higher in some regions than in others and the protective effect as well as the possible adverse effects of further layers of coding are controversial. The differences in attitudes could disappear to some extent if evidence based answers to at least some of the empirical questions were available. In the presence of uncertainty, the only way to reconcile different opinions might be to adopt the position of those who are the most risk-averse and to pay the additional costs for a more complex coding system. As a matter of fact, risk of errors in coding could probably be reduced in a more costly system using technical devices to check security and to maintain backups of all codes.

It is more difficult to resolve the controversy concerning different balancing of values used as an argument for and against different coding types. Respondents agreed largely on the necessity to protect confidentiality. The disagreement concerns the price different respondents are willing to pay for this protection in terms of admitting relative sacrifices of other values such as efficiency, quality of research and research participants' choices.

While technical solutions might exist to maintain efficiency and quality of research in spite of double coding, respect for participants' choices, in the absolute way in which it is defended by some interviewees from the US sample, might become incompatible with the positions on coding taken by most other respondents. It could be possible, although also costly, to respect the principle of "data availability" for research participants, meaning that they are guaranteed a right to access their own data at any point in time. However, letting each research participant choose his or her preferred type of coding seems impossible without sacrificing efficient functioning of most biobanks and without neglecting the security interests of particular groups of participants.

More research is needed on the feasibility of reconciling different values. Some respondents who hold extreme positions on the respect for participants' choices might modify their demands if they were confronted with the arguments of experts who have used or stored samples.

3.2. Contributions to the current debate

What do the results from our study contribute to the current debate and to the discussion about guidelines?

First, our results are useful because they provide a description of the range of opinions and the range of different contexts that have some influence on attitudes. When discussing ethically appropriate coding systems, guidelines should address the fact that international biobank research might have to adapt to local traditions concerning the role of physicians, researchers and trust.

Second, our study shows that in the case of a "typical" biobank described in scenario A, experts agree that irreversible anonymization is inadequate. In addition, as pointed out by respondents from our study, irreversible anonymization does not protect against abuse of information, if re-identification is possible through data linking and statistically rare characteristics (see above 2.2.3). Guidelines advising irreversible anonymization as the default standard such as UNESCO (2003) and WMA (2002) might therefore be misleading, both in terms of the protection provided by irreversible anonymization and its adequacy for typical biobanks.

Last but not least, we conclude that the ongoing debate should be enlarged and focus more extensively on the details of coding. The answers of the respondents show to which extent the judgments on coding systems depend on empirical questions. More empirical data are needed about the consequences of various types of coding, including the associated risks of errors and abuse. After having reviewed the arguments of the respondents, one is not surprised that guidelines are predominantly silent on details of coding. The present uncertainty about consequences makes it difficult to justify any detailed recommendation in favor or against a particular type of coding. However, instead of only asking for "confidentiality protection" or "data security," guidelines might be more helpful if they could provide more details on the criteria that contribute to the protection of both. Interviewees against double coding, and even many of those who supported the coding policy in scenario A.1, mentioned aspects of the coding system that should be added or specifically evaluated, in order to make the coding acceptable. The answers of the experts who participated in our interviews can be helpful to create a list of details that should be examined before adopting a coding system for biobank research. Such a list could be useful for institutional review boards, ethics committees and policy makers who will be confronted with the task to approve a coding system for a given or future biobank.

List of issues to consider when judging a coding system for biobank research

General	-	Make sure that the coding system is clearly understandable for all persons involved
	-	A particular coding system is necessary only if good reasons exist to use it
Protection of privacy/ confidentiality	-	The protection itself is important and the degree of protection must be evaluated with respect to various coding and repository types
	-	Evaluation of the protection should consider:
		1. the size of the region where samples stem from (a small region versus several countries, etc.)
		2. the type and details of personal information attached
		3. the risk of identification for cases that contain rare characteristics
		4. the possibility of using simple coding: not all samples and all types of research are always that controversial. Simple coding could be acceptable depending on the amount of clinical information attached to the samples
		5. whether number codes are sufficient or whether more than simple number codes are required, i.e., a higher degree of encryption
		6. how to ensure that, in particular, sensitive lifestyle information is protected
		7. the risk of abuse of information by third parties such as private insurance, health insurance
		8. the "recontacting risk." If individuals involved in the biobanking could be tempted, require more complex coding
Control instance for the coding	-	Exclude the existence of conflicts of interest for those involved in the control of coding, e.g. require a conflict of interest statement
	-	Define rules of agreement when it is appropriate to go back to the tissue donors
Quality of the research	-	Evaluate whether the coding affects the quality of research
	-	Evaluate and prevent possible risks of errors due to the coding
	-	Avoid "overkill" (unnecessarily complicated coding)
	-	Take into account the costs of the coding
Wishes of tissue donors and information	-	Evaluate the wishes of tissue donors concerning coding and inform them about the choices (or the absence of choices) they have with regard to the coding and about the benefits and risks of various forms of coding
	-	It should be clear to tissue donors under which circumstances codes will be released to permit recontacting
Biobanks involving samples from family members	-	Define the coding and feedback policy regarding those relatives who will not be the patients of the coding physicians
Involvement of physicians	-	Exclude the existence of conflicts of interest for physicians, e.g. require a conflict of interest statement

- Provide a mechanism for the handover of key-holder and all other biobank related responsibilities in the case of physician retirement
- Local physicians need training because they are not used to do coding
- It should be clear to physicians under which circumstances they have to keep the code confidential and in which cases they are allowed or even obligated to transmit information. Evaluate the usefulness of a contract with physicians which defines these circumstances.
- Establish guidelines on how physicians generate numbers for the coding: e.g. indicate that random numbers have to be chosen
- Define the conditions under which physicians keep the code (e.g. on secure computers without network access)

Bibliography

Abbott, A. (1999), "Sweden sets ethical standards for use of genetic "biobanks,"" *Nature* 400, 3.

Auray-Blais, C. and Patenaude, J. (2006), "A biobank management model applicable to biomedical research," *BMC Med Ethics* 7, E4.

Clayton, E.W. (2005), "Informed consent and biobanks," *J Law Med Ethics* 33, 15-21.

COE (2006a), *Recommendation Rec (2006) 4 of the Committee of Ministers to Member States on Research on Biological Materials of Human Origin* (Strasbourg, France: Council of Europe).

COE (2006b), *Draft Explanatory Memorandum for Recommendation Rec (2006) of the Committee of Ministers to Member States on Research on Biological Materials of Human Origin* <https://wcd.coe.int/ViewDoc.jsp?Ref=CM(2006)21&Sector=secCM&Language=lanEnglish&Ver=add&BackColorInternet=9999CC&BackColorIntranet=FFBB55&BackColorLogged=FFAC75>, accessed July 2007.

Elger, B. and Caplan, A. (2006), "Consent and anonymization in research involving biobanks," *EMBO Reports* 7, 661-6.

European Society for Human Genetics (2003), "Data storage and DNA banking for biomedical research: technical, social and ethical issues. Recommendations of the European Society of Human Genetics," *European Journal of Human Genetics* 11, Suppl. 2, S8-10.

European Agency for the Evaluation of Medicinal Products (2004), *Position Paper on Terminology in Pharmacogenetics* <http://www.emea.eu.int>, accessed July 2007.

Godard, B., Schmidtke, J., Cassiman, J.J. and Aymé, S. (2003), "Data storage and DNA banking for biomedical research: informed consent, confidentiality, quality issues, ownership, return of benefits. A professional perspective," *Eur J Hum Genet* 11, S88-122.

Grotzer, M.A., Shalaby, T. and Poledna, T. for the Swiss Pediatric Oncology Group (SPOG). (2003), "Establishment of the Swiss Paediatric Oncology Group (SPOG) Tumour Bank," *SIAK-SPOG* 2003, 180-84.

Hoeyer, K., Olofsson, B.O., Mjorndal, T. and Lynoe, N. (2004), "Informed consent and biobanks: a population-based study of attitudes towards tissue donation for genetic research," *Scand J Public Health* 32, 224-29.

HUGO [Human Genome Organisation] Ethics Committee (1998), *Statement on DNA Sampling: Control and Access* <http://www.hugo-international.org/Statement_ on_DNA_Sampling.htm>, accessed July 2007.

HUGO [Human Genome Organisation] Ethics Committee (2003), "Statement on Human Genomic Databases. December 2002," *Eubios Journal of Asian and International Bioethics* 13, 99.

Knoppers, B.M. and Chadwick, R. (2005), "Human genetic research: emerging trends in ethics," *Nat Rev Genet* 6, 75-9.

Knoppers, B.M. and Saginur, M. (2005), "The Babel of genetic data terminology," *Nat Biotechnol* 23, 925-7.

NBAC (1999), *Research involving Human Biological Materials: Ethical Issues and Policy Guidance, Report and Recommendations, vol. 1.* (Rockville: National Bioethics Advisory Commission).

OHRP (2004), *Guidance on Research Involving Coded Private Information or Biological Specimens* (Rockville, MD, USA: Office for Human Research Protections).

Quantin, C., Bouzelat, H., Allaert, F.A., Benhamiche, A.M., Faivre, J. and Dusserre, L. (1998), "How to ensure data security of an epidemiological follow-up: quality assessment of an anonymous record linkage procedure," *Int J Med Inform* 49, 117-22.

Swede, H., Stone, C.L. and Norwood, A.R. (2007), "National population-based biobanks for genetic research," *Genet Med* 9, 141-9.

UNESCO International Bioethics Committee (2003), *International Declaration on Human Genetic Data* (Paris, France: UNESCO).

World Health Organization (European Partnership on Patients' Rights and Citizens' Empowerment) (2003), *Genetic Databases, Assessing the Benefits and the Impact on Human Rights and Patient Rights*, Geneva.

World Medical Association (2000), *Declaration of Helsinki*, adopted by the 52nd World Medical Association General Assembly, Edinburgh, Scotland.

World Medical Association (2002), *The World Medical Association Declaration on Ethical Considerations Regarding Health Databases* (Ferney-Voltaire, France: World Medical Association).

Chapter 10

Informing Participants about Research Results

Andrea Boggio

1. Returning research results, participants' right to know, and the repository's duties

Human genetic databases are designed to gather and process large amounts of phenotypic and genotypic data to help researchers advance our knowledge of a variety of diseases. Population biobanks in particular are commonly set up with the promise of studying the relationship between genetic and environmental determinants of common diseases. Human genetic databases are therefore geared towards generating information that will or at least have some potential to improve the health of the studied population and its members.

That research has the desired goal to generate information that may help individuals is not new to debates in research ethics (Ravitsky 2006). Traditionally, there is a requirement to act on clinical trial results relevant to the health of participants. This is for instance the case when preliminary results of a drug trial are sufficiently promising that withholding the new drug from study participants in the control arm is no longer ethically justified. On the other hand, there is no duty to feed back results derived from fundamental research (Knoppers et al. 2006) and in genetic epidemiology (Beskow et al. 2001), because ordinarily they cannot be linked to identifiable participants. What is challenging about biobanks is that biobankers have the ability to link research findings to individual research participants and, in many cases, to participants' blood relatives—provided samples and data are not irreversibly anonymized at some point, which is rarely the case.[1] Further complexity is added by the fact that biobanks and population databases often explore gene/disease associations for common diseases. It is not unusual for such associations to appear promising in early studies, only to be disproved by further research. Consequently, making an assessment of their clinical significance for study participants is especially difficult.

Policies often focus on participants' rights and expectations, framing the issue in terms of right to know and not to know, or in terms of altruism and donation usually associated with participation in biomedical research. However, little attention is

---

1 Several respondents linked the discussions of feedback to the anonymization process, often suggesting that the ability to feed research results back to participants is a strong argument in favor of forms of reversible anonymization of samples and associated data.

paid to discussing any possible obligation that biobankers and researchers involved with large-scale repositories might have to convey research findings to participants. Based on these new technological possibilities, should biobanks be required to feed this information back to participants? To participants' blood relatives? What research findings? How? This chapter analyzes whether and to what extent research participants shall be informed of research findings. To explore this aspect of the disclosure of research results, respondents were asked to express their view with regard to the following policy:

> When specific gene mutations linked to an elevated risk of developing colon cancer are discovered, the repository will notify the physician who provided the coded sample and request that the participant be informed of the gene mutation and of its implications for his/her health.

The proposed policy stated that participants will be informed, through their physician, about the discovery of gene mutations that are linked to an elevated risk of developing colon cancer. The policy purposely did not address a number of further issues concerning returning results to participants—informed consent requirements; right not to know; what kind of information must be given; repository's obligation to ensure that the physician informs participant. This methodological approach offered the opportunity for respondents to raise issues not explicitly mentioned in the policy and scenario, and to propose amendments to the policy. These issues were also listed in the interview protocol so that the interviewer was in a position to prompt respondents, if needed.

2. Results

2.1 Should participants be informed?

First, respondents were asked to express an opinion regarding the adoption of the policy on feedback proposed in the scenario. From the analysis, a substantial support for the policy emerges. The justification was primarily found in the idea that a participant has a right to know all information that may be instrumental to improve a participant's health. However, respondents also noted that the policy as described is not acceptable, because a provision ought to be added in the informed consent discussing the possibility of feedback. It is of utmost importance that these issues are raised in the informed consent so that participants are aware that they may be contacted with information in the future, respondents argued. In other words, research results cannot be fed back in the absence of participants' advance consent to the possibility of feedback. It is also reasonable to note that respondents were inclined to support the policy because the research findings at stake were *linked* to an *elevated risk* of developing colon cancer. The findings' rather clear clinical significance has certainly shaped respondents' confidence in the proposed policy. However, a (small) number of respondents disagreed on various grounds. Firstly, there was some disagreement on whether the informed consent is acceptable if no opportunity to decline receiving feedback is provided. This scenario assumes that research participants who are interested in being part of a study must also agree to the disclosure of research findings. The disagreement concerned whether participants have a right not to know. To some, it is a moot question as long as the informed consent form spells out clearly the terms of the arrangement. To other, an informed consent that does not include a provision recognizing that participants' right not to know is invalid.

Secondly, some respondents were not comfortable with the policy because participants are not really benefiting from these findings that, in their view, lack clinical relevance since removing the genetic risk of developing cancer is impossible. Lifestyle changes and increased screening were considered to be insufficient grounds to impose the feedback requirement on researchers. Some respondents, especially genetic epidemiologists, challenged the whole idea that repositories set up in the way described in the scenario could generate the kinds of individualized research findings described in the policy.

Thirdly, some argued that biobankers and repositories are primarily responsible for conducting genetic research and not to treat participants. A duty to disclose research findings goes beyond the scope of conducting biomedical research. Such a requirement would be too demanding, and it could result in a disincentive for researchers to conduct studies with possible clinical implications—an outcome that ought to be rejected on utilitarian grounds, respondents argued.

Fourthly, some argued that participation in research ought to be based on altruism, and research participants must not expect to receive any form of individualized feedback. "Biobanks should appeal to altruistic motivations on the part of the participants" (#13 (a) IS, No. Am./Asia, philosophy). Nevertheless, it may turn out that research subjects do expect that their participation in studies leads to acquiring information directly relevant to them, and that fact may be an incentive to participate in research based on non-altruistic motives.

Finally, others argued that the feedback policy has not appropriately expressed the view that feedback may distress participants, and ultimately harm them. The number of respondents who envisioned the possibility of harming participants by returning results was surprisingly large. Respondents were concerned that research participants might misinterpret the information. In fact, respondents from various regional backgrounds were concerned that participants may not be able to appreciate the weight to be given to research findings that are often not validated in more extensive studies. Overestimating these results could lead to "unnecessary alarm" (#82 IS, Europe, natural science, A/R), several argued:

> We are dealing with hypothesis, so it is tough to say. We need to be sure of the results ... It confuses the participant. If we tell them, and we are still in the research, we have to think about this information very carefully ... people like scientific clear results but they are not interested in the delicate, hypothetical structure of science ... If we tell people about these research results, we have to be sure that professionals deal with these results (#67 IS, Asia, genetics, government, biobank, R/A).

> ... a research study that sees an association if it is followed by two other studies that do not see an association is just going to drive the average patient crazy, which I worry

about a lot. I fear too much confusion (#98 US, philosophy, medicine, genetics, university, R/A/S).

It is unclear whether respondents' motives can be explained exclusively in terms of non-maleficence or whether the explanation lies beyond that—perhaps on paternalism or on the desire to avoid responsibility. A harm-based argument against sharing research findings back to participants has been seen in the potential harm associated with the inadvertent discovery that some of the family members involved in the process are not biologically related:

> Just imagine you belong to a family that had a cancer and then you find out that you weren't really [blood related], that you did not belong to that family, and therefore all your children who will come will be free of that. I think you should know that (#37 IS, Oceania, social sciences, bioethics, university, A).

Liability avoidance seems to have played some role in shaping the responses of interviewees from the US samples. Many construed the role of the repository in narrow terms, so that biobanks have no duty to act upon results with clinical validity. It is up to the physician to do something about them and decide whether and how to share them with any patient who participated in the study. The reluctance to accept any responsibility relating to returning results may be caused by concerns about legal liability or concerns about the responsibility of providing this unpleasant information.

The role of internal ethics committees has been stressed by a number of respondents who felt that decisions concerning sharing information with participants require a careful assessment of their validity and of the impact of its disclosure to participants.

2.2 What information must be fed back to a participant?

As regards the kind of information that must be conveyed to participants, the interview was set up so that respondents were prompted to take a position on various kinds of information—mutation related to other forms of cancers and/or to other health conditions; information of no clinical value but relevant to life planning, reproductive choices, or paternity determinations.

As in other cases, respondents were divided when it came to deciding about findings that were unrelated with colon cancer or that could not improve the clinical management of the disease. The respondents who favored a more inclusive approach were inclined to extend feedback to *all* kinds of information (with the exclusion of paternity-related information). However, the responses often focused more on the quality of the information passed on to participants rather than on its nature. In fact, while respondents were usually comfortable with various feedback arrangements— provided participants were explained the details of the feedback and had consented to it beforehand—many expressed concerns regarding the medical or clinical value of scientific findings that are, by their very nature, provisional evidence of links between genes and conditions, that are subject to further scrutiny and possible disproof.

As an alternative, or addition, to individualized feedback, several respondents suggested that repositories should make research findings known to participants in general terms by publishing newsletters or posting the information on websites. Besides providing beneficial information to participants, this arrangement was favored because it fits the view that participants and repositories ought to have an ongoing relationship of which a mutual exchange of information represents an integral part. Also those who argued that informed consent should be seen as a process felt that being updated on research developments and on new research projects is required for the consent to be valid over time.

Finally, as a general trend, respondents argued that sharing findings relevant to paternity as particularly problematic, often because repositories must use their resources to conduct biomedical research and not test for paternity. Very few respondents concluded that disclosure of this information was appropriate.

2.3 Who should be responsible for returning research results?

If research results are returned, then either the repository should pass on the information to physicians or it should contact participants directly. We drafted a policy whereby the information would be given to physicians with the recommendation to share it with the participant. We asked whether that was an acceptable arrangement and whether, in case a physician failed to share the results with his or her patient, the repository was responsible for contacting the participant directly.

The majority of respondents rejected this scenario involving direct contact with participants. Respondents argued that physicians are in a better position than researchers to assess how the sharing of information will affect the participant: "it is easier to contact physicians, they understand implications of data and they are able to provide counseling" (#29 US, law, government/university, O). Therefore physicians ought to be the exclusive source of feedback for research participants. We have seen earlier how avoiding liability could have been a pressing concern to many respondents, especially among US respondents. Others argued that physicians are primarily responsible for returning results, because providing direct, individualized feedback exceeds the scope of activities of a biobank.

Other respondents were less satisfied with the mechanism set forth by the policy, arguing that family doctors may not be sufficiently trained to both understand and communicate to patients the complexities of genetic information. Based on this assumption, they argued that scientists are more appropriate to transfer genetic information, which by its very nature is complex and often tentative at this stage of research, to participants. The concern that "the individual physicians involved in the study would not be so sophisticated about the information as to be able to fully interpret it" (#36 US, medicine, genetics, government, R/A/S) was therefore raised by several respondents.

Others expressed no preference on who should be responsible for communicating the findings provided (i) that the feedback arrangement is explained and understood by the participants from the outset and (ii) that people with expertise in genetic counselling become involved when the findings become available. Therefore, to many respondents it is important that appropriate genetic counselling expertise is available at the time the sharing of the research findings takes place.

3. Analysis and implications

Whether biobanks should return research results remains a controversial issue. On one hand, the concern that information that can be helpful for clinical or non-clinical purposes (such as planning life or reproductive choices) may not reach research participants is perceived as increasingly problematic. On the other hand, the uncertainties associated with assessing genetic risks and findings generated by fundamental research, especially when not yet fully validated, raised concerns among respondents. Part of the problem is that technological advances are blurring the line that used to divide clinical practice from scientific research: population-based databases aiming at studying common diseases may generate information that could be easily linked to an individual participant by overriding the code concealing his or her identity. The implications for public health and for the role of researchers are significant, and largely unexplored. However, if compared to the responses in other areas explored in this study (for examples, property rights in DNA and benefit-sharing), respondents showed a substantial degree of familiarity with the major debates in this area. Therefore, the responses were more coherent and articulate than those collected on other questions.

Traditionally, policies and scholars have analyzed the issue in terms of participants' right to know and right not to know, and as a consequence, great emphasis has been placed on the role of informed consent. The data of this study is consistent with this perspective. However, this approach could be complemented—as suggested by a substantial number of respondents—with disseminating to the public information about the research results by posting them on the repository's website, by mailing newsletters, and by sending to participants individualized letters describing the results in general terms.

As previously discussed, to envision a feedback of research results entails revisiting the role of researchers, and their obligations towards research participants. Traditionally, researchers have not been required to share research findings with participants, except indirectly, as in the case mentioned earlier where equipoise is broken in a clinical trail and commands a new clinical course of action for some participants. However, biobanking often creates an ongoing relationship with participants, in which scientific hypotheses that were not contemplated when samples were collected are pursued over time, in which new health and personal data regarding individual participants are added to the database, and in which the possibility of linking individual findings to individual participants remains open. One may reasonably ask whether the traditional paradigm built upon a clear separation between research and clinical aspects can survive the new developments (Fernandez et al 2003; Fernandez et al. 2004). A substantial number of respondents were willing to impose obligations on researchers beyond the mere creation of general knowledge. To them, beneficence requires that information potentially of use to participants ought to be passed on to them, either indirectly through physicians or directly from researchers with the support of genetic counsellors. This trend is counteracted by the view—also popular among respondents—that returning research results is beyond a biobank's duty and/or that sharing results could do more harm than good.

This is also an area of the study where differences based on the professional and/or regional background of the respondent clearly mattered. Respondents from regions with poor health care systems and limited training opportunities for doctors tended to see the repository as ultimately responsible for handling genetic information. To these respondents, researchers involved with the repository offered better prospects of understanding the implications that research findings might have for participants' lives and of providing feedback in a manner that would not harm the recipient. On the other hand, respondents from regions where health care systems are of comparatively good quality, indicated that genetic counselling is beyond the scope of conducting research and that clinicians with proper training in genetic counselling ought to be responsible for conveying research findings back to participants. A simple lesson is that we cannot take policies out of the context in which they will operate. Conducting research in regions where basic health services are inadequate, capacity for genetic counseling limited, and access to medical journals difficult, could bring an invaluable contribution to the communities involved. In fact, researchers would be in a position to offer training opportunities for local doctors and services to individuals that would ordinarily be unavailable. Arguably, such need is less compelling in wealthier countries, with more effective health systems. However, technology offers new opportunities for making links between researchers and participants. Personally controlled health records and the "informed cohort" model, proposed by Kohane and colleagues, allow researchers to communicate with participants without violating anonymity, while giving subjects control over both the types of information they submit and the information they receive (Kohane 2007).

To conclude, the debate on biobanking and the sharing of research results challenges the traditional view that researchers are not required to share research findings with participants. Implementing participants' right to know is controversial, and researchers may be required to re-think their relationship with participants, and more generally with the public. This is a case in point that advances in technology raise new challenges and perhaps new ethical obligations.

Bibliography

Beskow, L.M. et al. (2001), "Informed consent for population-based research involving genetics," *JAMA* 286, 2315-21.

Fernandez, C.V., Kodish, E. and Weijer, C. (2003), "Informing study participants of research results: an ethical imperative," *IRB* 25, 12-19.

Fernandez, C.V., Skedgel, C. and Weijer, C. (2004), "Considerations and costs of disclosing study findings to research participants," *CMAJ* 170, 1417-19.

Knoppers, B.M., Joly, Y., Simard, J. and Durocher, F. (2006), "The emergence of an ethical duty to disclose genetic research results: international perspectives," *Eur J Hum Genet* 14, 1170-78.

Kohane, I.S. et al. (2007), "Reestablishing the researcher-patient compact," *Science* 316, 836-37.

Ravitsky, V. and Wilfond, B.S. (2006), "Disclosing individual genetic results to research participants," *Am J Bioethics* 6, 8-17.

Chapter 11

Ownership of Samples and Data and Territorial Restrictions Concerning Data and Samples beyond National Boundaries

Andrea Boggio

1. Researching ownership of samples and data

Ownership rights regarding genetic material are controversial. New developments in genetic research and biotechnology have certainly revived the traditional debate on body ownership and its commodification (Charo 2006). While a few drops of blood are themselves of little scientific or economic value, a collection of biological material from a well documented cohort of donors has an aggregate, scientific and economic value that a sample alone does not possess. Indeed, the value is even greater if linked to medical records and other data regarding the donor of the sample sources are linked to the material. Therefore, biobanking gives value to goods that *per se* amount to little worth, and that produces contested ownership rights.

Biobanking involves balancing the sometimes conflicting interests of different stakeholders: donors/research participants, biobankers, investigators, and the public. Ownership is one of the areas in which the balancing of interests is particularly problematic around at least two sets of stakeholders' interests that are in tension.

The first tension arises between research participants and biobankers. Part of the tension arises because of the inconsistent way legal interests in the human body are characterized when possessed by individuals (in their own body) as opposed to when possessed by a third party outside the body. Typically, individuals participate in biomedical research for altruistic reasons, and any biological material taken from them is treated as a donation. Legally, that entails the donor having no claim to the biological material after it is donated. Once biobanks collect, store, and organize the donated material, particularly by linking it with related data, the biobanked samples undeniably acquire value. Not surprisingly, biobanks therefore assert an ownership interest in the donated materials (and related data), even though legal systems would not generally assign an ownership interest in their own bodies to the individuals who had donated the materials in the first place. Clarifying who may assert what control over biobanked samples, and whether that control should be effectuated by recognizing property rights, are therefore important legal and ethical questions. Should ownership rights be assigned to the biobank? Do the people who participate in research by donating biological materials to a biobank have an ownership claim regarding such materials? Who should be able to control the samples when a biobanks and a participant/donor disagree?

The second tension involves nation states. In recent years, several national biobanks have been established to store samples and data exclusively from people living in particular countries. Some of the countries—including Estonia, Latvia, Sweden, and Brazil—have laws that restrict the circulation of samples and data outside the country in which the biobank is established. In Israel, the transfer of the entire national database to foreign soil is prohibited. On the other hand, genomics research has become increasingly global and relies on cross-boundaries transfer of material and data (Muula 2007). Large-scale sequencing projects—such as the HapMap (Integrating Ethics and Science in the International Hapmap Project 2004) or the Welcome Trust-funded effort to understand the genomics of malaria, which involve investigators from a large number of research sites—would not be possible if samples could not be transferred from one country to another. Moreover, international bodies encourage cross-national exchange of tissue and data. Article 18 of UNESCO's *International Declaration of Human Genetic Data* provides that:

> States should regulate … the cross-border flow of human genetic data, human proteomic data and biological samples so as to foster international medical and scientific cooperation and ensure fair access to this data.

Thus, a tension clearly exists between the interests of countries that are trying to exploit the human genomic resources assembled within defined geographical and political boundaries, on the one hand, and transnational efforts to use samples and data from different regions, on the other hand. Should countries have a moral right to impose restrictions on the transnational circulation of samples and data? Or should a general duty be recognized to share samples and data with investigators of other nations?

Other issues associated with ownership and circulation are discussed in other subchapters of this book (including commercial transactions involving human biological samples, transfer of samples, destruction of samples, and intellectual property rights). In this subchapter, we present findings regarding two questions: who should be able to assert what ownership rights over stored human biological samples, and what territorial limitations concerning the circulation of data and samples beyond national boundaries are appropriate? To address ownership issues, we first looked at policies: whether biobanks ought to be named the owner or custodian of stored samples and resulting data; during the interview, we also explored alternative arrangements—in particular the ownership being assigned to research participants—and the practical implications of the various policy arrangements. To address ownership issues, Scenario B presented a non-profit biobank faced with a proposed government policy under which it would not be permitted to transfer samples and associated data out of the country. We asked respondents whether they agreed with this policy arrangement and, to stimulate and organize their responses, we prompted them with three arguments in favor or against circulation. The first argument was the most theoretical, namely that, since the human genome is part of the common heritage of humanity, such limitations are unacceptable. The second suggested that territorial restrictions are justified as a way of building up capacity, by encouraging researchers to come to the country where the biobank is located to do their work. Finally, we cited the argument that territorial limitations are justified based on the fear that a country cannot protect the confidentiality of information about participants once samples leave the country.

2. Results

2.1 Biobanks: custodians or owners?

Ownership relates to the control over "things," and it is commonly associated with the ability to use, control, transfer, or otherwise enjoy the owned "thing" (Waldron 2004). Access and control are therefore associated with various rights—the right to exclusive possession, to manage use by others, to the income from use by others, to the capital value, including alienation, consumption, waste, or destruction, and other rights. To capture the complexity of the rights associated with the concept of ownership, property has been famously described as a "bundle of rights" (Honoré 1961).

Since the advent of biological and epidemiological studies involving stored human tissues, legal systems and ethicists around the world have had to struggle with questions of ownership of human tissue and genetic data. Policy recommendations and legislative approaches reflect the difficulty in agreeing on the ownership of biological samples once they have been removed from the human body. Discussing these issues in the context of biobanks raises questions not only of the property-characteristics of the collected samples themselves but of the genetic data derived from them and of the information that resides in the biobank and that collectively makes up the biobank, which is an organized set of information (consisting of biological and digital material). Not surprisingly, these questions lead in turn to questions about the permissibility of commercial transactions involving human biological samples.

For the purpose of the present study, the ownership of the samples stored in a biobank was addressed by asking whether the repository should be the owner or the custodian of the samples stored (a custodian being someone who has charge of property on behalf of someone else). The majority of respondents from both the international sample and the US sample were in favor of the proposed policy providing that the repository is the custodian and not the owner of the samples. Respondents reached this conclusion based on two fundamental reasons: the dissatisfaction with the idea of ownership and the assumption that research participants own the samples that have been removed from their bodies and donated to the biobank.

First, respondents favored custodianship because of general dissatisfaction with the idea of ownership of human samples. Respondents referred to ownership of samples as being "too strong" (#30 IS, Europe, medicine, bioethics, government, R/A/S), "tricky" (#42 IS, Asia and No. Am./Asia, medicine, bioethics, government, A; #68 IS, No. Am./Europe (nationality: No.Am.), life sciences, consultancy R/A), "too theoretical" (#57 IS, No. Am./Europe, medicine, university, S) or prompting the idea of "irreversibility" of the legal status of samples entailed in a transfer of ownership (#53 IS, Europe, medicine, public health, Int. Org., R/A/S). It is unclear, however, to what extent respondents had sufficient familiarity with the legal distinction between ownership and custodianship. Most of the respondents were not trained in law, and failed to focus on the legal implications of ownership. This may be a problem especially when drafters of policies or international statements use the words in a precise fashion that does not take account of their more casual or even metaphorical use in non-legal circles.

The most confusing aspect for respondents was the relationship between donors' being able to withdraw their samples from the biobank and the bank's status as owner or custodian of the samples for the public good. Few respondents were able to articulate this relationship with distinction among the various "rights" that are associated with ownership (such as those as defined by Honoré). Some respondents— those mostly concerned with donors' rights—feared that assigning ownership to the biobank would result in a substantial barrier to the destruction of samples and data if the donor decided to withdraw consent.

Second, respondents reasoned that custodianship is the appropriate arrangement because each participant owns his or her sample even after it is stored in a biobank. The respondents who favored this view can be classified in two categories. The first group comprises "biobankers," that is, respondents who are involved with organizations storing samples and who are unlikely to be principal investigators of any of the research projects using the samples stored in their biobank. Several biobankers construed their professional role as "mediator" (#30 IS, Europe, medicine, bioethics, government, R/A/S) between research participants and scientific investigators, and concluded that research participants retain property rights in the stored samples. This viewpoint is complemented by the idea that the biobank acts as the custodian of the samples. Custodianship was considered to be an attractive concept enabling mediation among competing interests without sacrificing the biobank's "right to hold the material" (#68 IS, No. Am./Europe (nationality: No. Am.), life sciences, consultancy R/A). The second group is comprised of respondents that were selected because they expressed an indigenous group's viewpoint. They consistently argued that human DNA ought to be considered the property of the research participants. These responses are certainly consistent with indigenous peoples' claim that the genetic resources that originate from members of such groups are a collective possession and DNA as defined by some indigenous cultures "is more comprehensive of a description of [its] biological elements [because it also includes the concepts of] genealogy and the spirit of men" (#14 IS, Oceania, political science, indigenous IP, university A/R; see also, Reynolds 2007; Paoakalani Declaration Ka "Aha Pono 2003). In fact, many indigenous peoples have claimed that their genetic material is inalienable and is to be protected as part of their cultural heritage under the international instruments recognising indigenous groups' rights, such as the United Nations Declaration on the Rights of Indigenous Peoples (Harris and Kanehe 2006). Many indigenous groups have in fact "identified a cultural and spiritual relationship with genetic material when examined through a cultural lens" (Harris and Kanehe 2006). Finally, some respondents developed this idea even further by suggesting that sample ownership should be a collective one. Under this view, the samples belong to the group that participates in research as a collective entity or to the family of which the individual participant is a member.

On the other hand, a substantial minority, especially in the US sample, said that the depository should own the sample. Clear recognition of a legal right would be best for practical purposes, to prevent abuse, and to entitle the biobank to protect the collection. Respondents also argued that, if the depository were not the owner, then it would be nearly impossible for researchers or the depository to do their work— "it gives the impression that people can come back and take their tissue back out" (#98 US, philosophy, medicine, genetics, medicine, university, R/A/S). Respondents holding this position usually added that research participants must be told in the informed consent document that the biobank will acquire ownership of the samples collected from them.

Discussing ownership certainly requires a substantial familiarity with the legal concept of property, which many respondents did not possess. This raises the issue of how scientists understand legal concepts and what use they make of them. Most respondents construed and analyzed ownership in terms of relationship between participants and biobanks. To some, ownership should be assigned to the research participants to protect them against abuse on the part of biobanks or investigators. To others, ownership should be assigned to the biobank because stored samples are public goods, and biobanks can only operate if they have full title over the stored samples. For these respondents it was still essential that participants be informed and consent to the transfer of ownership; they saw nothing inconsistent with a biobank being the owner and participants still being offered a reasonable opportunity to withdraw consent.

2.2. Territorial restrictions concerning data and samples beyond national boundaries

Restricting the international circulation of samples and associated information was seen as ethically unacceptable by many respondents. Respondents substantially sided with the argument that territorial limitations are ethically dubious or, at best, acceptable but not desirable.

Two arguments were most commonly cited. First, to many, science is global—as one interviewee noted, "we should work in a global way" (#30 IS, Europe, medicine, bioethics, government, R/A/S): scientific research is a worldwide effort involving all scientists worldwide; to advance, science needs sharing; and therefore no policy should prohibit cross-national circulation of genetic material. Indeed limitations are "contradictory with [the] idea that … research is a public good" (#24 IS, Europe, medicine, consultancy, A/R). Second, although nothing prevents countries from adopting restrictive policies, sharing favors scientific research and territorial limitations "would destroy the scientific usefulness of the samples" (#32 IS, Europe, medicine, university, S/A) because "international comparison is an essential part of scientific research" (#67 IS, Asia, genetics, government, biobank, R/A). This view was particularly popular among respondents from the US sample, possibly because a great number of repositories and research institutions in the United States store or work with samples that come from the four corners of the world. Consequently, US respondents seem to value international collaborations highly.

By contrast, other respondents argued that the policy restricting sample and data circulation is ethically permissible. First, some argued that it is the duty of a biobank to protect participants. Given the comparative diversity in legal standards and mechanisms to protect participants, preventing samples from coming under the legal regimen of a foreign country is seen by some as a reasonable means of protecting participants. Second, others argued that, as part of setting their domestic policy agendas, countries are free to choose the policy arrangement that best fits their interest (including imposing territorial restrictions), and these choices must be respected by other countries. Third, others argued that genetic heritage can be seen as a national resource along with oil, fisheries, and others. Consequently, just as countries cannot be forced to share national resources, restrictions on access to or export of their genetic resources may also be appropriate:

> I consider them as DNA heritage of the country and it is the only agent entitled to control the fate of this information (#93 IS, Europe/Middle East, biology, genetics, biobank, S).

Fourth, restrictions were seen as appropriate if they aim to protect national security. In particular some international sample respondents argued that it is possible that genetic information about an entire population may be used by bioterrorists to plan terrorist attacks against that populations (62 IS, Europe and No. Am/Europe, medicine, law, university, R; 65 So. Am., life sciences, genetics, university, A; #93 IS, Europe/Middle East, biology, genetics, biobank, S). Interestingly, no US sample respondent raised this point.

However, the question of confidentiality was less problematic than expected. Although the need to protect confidentiality is "always an issue" (#5 IS, No. Am./Europe, medicine, university, S), confidentiality must be respected, and its enforceability depends very much on the context rather than on abstract, possible scenarios. In fact, respondents said, it depends on what "agreement is signed" when samples leave a country (#30 IS, Europe, medicine, bioethics, government, R/A/S), on the "need for national and international policies to be in harmony as much as possible" (#46 IS, No. Am., law, social science, NGO, A), on the "safeguards that are in place" (#73 IS, No. Am./Asia, life science, government, S), and on whether "the foreign researcher follows the guidelines" (#94 IS, Asia, life sciences, government, A). However, respondents suggested that one must not presume that confidentiality will be at risk just by the mere fact that samples are transferred outside the national boundaries:

> [I] … it is a problem but some places outside might have better security than the University next door … [B]eing in different countries is less relevant than the quality [of the safeguards] (#68 IS, No. Am./Europe (nationality: No. Am.), life sciences, consultancy R/A).

Finally, territorial restrictions aiming to protect the investment required to undertake large genetic studies are justifiable:

[I]f you consider small groups and in particular specific populations in specific regions, then [imposing territorial limitations] might be the only way to reasonably start some kind of collection [in terms of public acceptance] … [I]in large collections you have to invest money to create the value [then your limitations are acceptable] (#24 IS, Europe, medicine, consultancy, A/R).

To others the idea of territorial limitations is an illusion in practical terms, especially when private companies are involved with the research:

When people collected those samples, they thought they had to keep the samples in [this country], but the study was supported by pharmaceutical companies [which] are so international, meaning that the information is not kept forever in [the country where samples are collected]. So it is meaningless to keep this kind of policy (#67 IS, Asia, genetics, government, biobank, R/A).

Respondents were also prompted to discuss the argument that, since "the human genome is part of the common heritage of humanity," territorial limitations on the circulation of samples and associated information are unacceptable. Although some respondents found the "common heritage" concept somehow related to the issue, the majority of experts—both those against and those in favor of territorial limitations— found it either weak or irrelevant. Those who found the common heritage argument weak were mostly concerned that the idea of common heritage is "a blurred concept" (#06 IS, So. Am, philosophy, bioethics, university, A), that is "non conclusive" (#48 IS, No. Am./Europe, philosophy, university, R/A), and, even if the contrary were true, a respondent pointed out that:

[T]he actual practice should be understood in a much more realistic way. If you stick to the principle too much, nothing can be done (#94 IS, Asia, life sciences, government, A).

Those who viewed the common heritage argument as irrelevant argued that other considerations (especially participants' protection) were more important:

Just because it is part of the common heritage of humanity, it doesn't mean that it is open to abuse from anyone who wants to make any use of them (laughing) (#82 IS, Europe, natural science, A/R).

[The argument] conflicts with the notion of the integrity of the persons and their right of making decisions on the use of their biological material. Basic human rights come first (#46 IS, No. Am., law, social science, NGO, A).

[A]s far as the genome is concerned, there should be the common treasure. But the use of data of human beings of course is another issue (#73 IS, No. Am./Asia, life science, government, S).

The last word [on the fate of the sample] has to be with the donor. If it is OK that 'my sample is for research in my country,' that has to be respected (#17 IS, Europe, law, university R/A).

[W]hat are they going to use it for, once it becomes universal? (#37 IS, Oceania, social sciences, bioethics, university, A).

[T]o say 'your DNA is like my DNA,' so what's so great about it? ... [B] but there are specific features in some populations. ... [T]there are questions about developed and developing countries and countries which are rich in biodiversity (#42 IS, Asia and No. Am./Asia, medicine, bioethics, government, A).

3. Conclusions

In this chapter, the empirical findings regarding three questions raised by the ownership, and restriction to the circulation, of genetic material are discussed. The data show that the respondents struggled with the question of sample ownership. Although the idea of ownership is certainly challenging to policymakers, some consensus was reached that what is important is to reason about the specific duties that investigators owe to the participants and the public rather than discussing abstract legal categories. In this perspective, the custodianship model was more appealing to many respondents because of its flexibility and its focus on duties as opposed to rights. Interestingly, demands for legal certainty were voiced more strongly by investigators (biobankers in our samples) and by vulnerable populations (indigenous peoples in our samples).

Regarding territorial restriction, reasonable arguments for and against were offered by the respondent: to some, restrictions are not desirable; to others, restrictions are permissible (and inevitable). Common ground can be found when considering that ultimately respondents did not disagree on the basics. They looked at the same glass: however, while some saw it empty, others saw it full. In other words, in an age of increasing global exchange of ideas, territorial restrictions may look a surprising (if not anachronistic) strategy to protect something—a national, genetic treasure—that by its very nature defies geographical containment. Yet, restrictions do not seem to violate fundamental ethical principles but only the ethical aspiration that scientific research ought to conducted in the common interest of humanity.

Bibliography

[No authors listed] (2004), "Integrating ethics and science in the international Hapmap Project," *Nature Reviews Genetics* 5, no. 6: 467-75.

Charo, R. Alta (2006), "Body of research—ownership and use of human tissue," *N Engl J Med* 355, no. 15: 1517-19.

Harris, D. and L.M. Kanehe (2006), "Asserting tribal sovereignty over cultural property: moving towards protection of genetic material and indigenous knowledge," *Seattle J. for Soc. Just* 5: 27-55.

Honoré, A.M. (1961), "Ownership," In *Oxford Essays in Jurisprudence: First Series,* edited by A.D. Guest (Oxford: Clarendon Press), pp. 107-28.

Muula, A.S. and J.M. Mfutso-Bengo (2007), "Responsibilities and obligations of using human research specimens transported across national boundaries," *J Med Ethics* 33, no. 1: 35-8.

Paoakalani Declaration Ka "Aha Pono. Paper presented at the Native Hawaiian Intellectual Property Rights Conference, Waikiki, Hawai'i

(October 2003), Available at <http://www.ilio.org/Ilioonline/ahapono04/Paoakalani%20Declaration.pdf>, accessed 19 April 2007.

Reynolds, P. (2007), "The sanctity and respect for Whakapapa: The case of Ngati Wairere and Agresearch," in *Pacific Genes and Life Patents, Pacific Experiences and Analysis of the Commodification and Ownership of Life*, edited by Aroha Mead and Steven Ratuva, Call of the Earth [Llamado de la Tierra] and The United Nations University Institute of Advanced Studies.

Waldron, J. (2004), "Property," in *The Stanford Encyclopedia of Philosophy*, edited by Edward N. Zalta. Available at <http://plato.stanford.edu/archives/fall2004/entries/property/, accessed 20 April 2007.

Chapter 12

Public Domain Sharing, Patents, and Fees Resulting from Research Involving Genetic Databases

Andrea Boggio

The commercialization of genetic research is growing. Knowledge generated by fundamental research is increasingly translated into commercial products offered for sale to the public. The legal and ethical debate on the commercialization of genetics research is very polarized. One reason is that general worldviews are in conflict as to the interplay of commercial interests and the public interest, and this is particularly true as regards health care. More specifically, opportunities for commercialization are seen by some as difficult to reconcile with the interests of participants in research. It is argued that the altruistic nature of participation is inconsistent with the opportunity for downstream users to commercially exploit knowledge that would not have been produced without research subjects' altruistic willingness to take part in studies. Moreover, while researchers are rewarded with intellectual property rights (IPRs), ordinarily research participants hold no property rights (as discussed earlier in the chapter on the ownership of samples and data). Finally, ethicists and policymakers often claim that all forms of monetary compensation for participation in research are inadmissible, while commercial exploitation of participation is not only acceptable but desirable, because it provides an incentive to do research and develop new products and techniques. In addressing the commercialization debate, the present study focused on three areas: the obligation to put data in the public domain; patenting rights and publicly funded research; and, the admissibility of fees imposed on researchers using a repository.

1. Researching commercialization: public domain, patents, and fees

Legal systems routinely protect and reward inventors for their creativity and ingenuity. A patent gives an inventor the exclusive right to make, use, and sell an invention for a limited period of time. To qualify, an applicant must satisfy certain requirements, which vary across jurisdictions. In the United States, patents are awarded only if the invention, discovery, process, or design is genuine, novel, useful, and not obvious in light of current technology. The legitimacy of patenting genes has been debated for a long time (Watts 2007). Genes are "products of nature," and therefore they are not supposed to be subject to patent law because they fail to meet the "non obviousness" requirements (*In re O'Farrell*, 853 F2d 894 (Fed Cir. 1988)). Yet, patents for genetic inventions have been routinely awarded in the Unites States, in Europe, and elsewhere (European Parliament and Council 1998). Research shows that "nearly 20% of human genes are explicitly claimed as U.S. IP" (Jensen and Murray 2005). Jensen and Murray add that this figure represents 4,382 of the 23,688 of genes in the NCBI's gene database as of 2005; roughly 63% are assigned to private firms; and, the top assignees are all US-based companies (Jensen and Murray 2005). More recent data show that companies are applying for fewer gene patents and that they are actively preventing them by putting gene mutations databases in the public domain (Hopkins et al. 2007). A further complication results from the claim that, without awarding patents, researchers and companies would lack an incentive to engage in scientific research and develop new products. However, after a thorough review of the existing support of this claim, John Barton recently concluded that "[e]mpirical evidence on the role of intellectual property protection in promoting innovation and growth remains inconclusive" (Barton 2007).

Gene patenting is also controversial in debates about population genetics research with vulnerable populations. Since the 1990s, several research initiatives have raised controversies, especially when involving traditional knowledge and indigenous resources. This was the case for the decision of the National Institutes of Health (NIH) to patent a cell line found in the blood of a man from the Hagahai tribe of Papua New Guinea (Riordan 1995), the "biopiracy" critique launched against the Human Genome Diversity Project (South and Meso American Indian Rights Center 1995), and the opposition against the efforts of a part of an Australia-based biotechnology company to establish a database in Tonga (Burton 2002).

An alternative to a property-based approach emphasizes the public domain, namely an arrangement under which no person or other legal entity can establish or maintain proprietary interests within a particular legal jurisdiction. If data and other information are put in the public domain, then the public owns them, rather than an individual researcher or institution. In addition, anyone may use or exploit them, whether for commercial or non-commercial purposes. Increasingly, the notion that genetic data should be put in the public domain is gaining recognition. Recently, Lowrance and Collins noted that "[a] cultural habit of rapid, open release of genomic data has been pursued by the involved scientists and institutions since the beginning of the Human Genome Project" (Lowrance and Collins 2007). All genome sequences generated by the Human Genome Project have been deposited into GenBank, a public database freely accessible by anyone with a connection to the Internet. Similarly, all single nucleotide polymorphisms (SNPs) mapped by The International SNP Map Working Group were put in the public domain (Marth 2001). Parallel to the arguments raised in the IPRs debate, supporters of the public domain model argue that it offers more efficient dissemination of information, encourages its widespread use, minimizes transaction costs, and makes R&D cheaper and faster (Rai and Boyle 2007). Based on the growing culture of sharing among scientists, Foray argues that an "open science model" based on an IPR free-zone is emerging, countering the still dominant commercialization model (Foray 2004).

Finally, concerns about commercialization relate to the ability of a biobank to exploit its collection for commercial purposes. Before the biobank era, researchers could simultaneously collect samples, conduct research, and exploit the commercial benefits—mostly in the form of royalties. Biobanks and similar genetic databases have challenged this model because they are research infrastructures that collect and store samples, that sometimes perform fundamental research, but that rarely conduct the kind of (applied) research that leads to commercial gains. While it is unlikely that they benefit themselves from the profitable exploitation of their holdings, biobanks are valuable yet costly resources. Commentators agree that biobanks may request external researchers to pay fees to cover expenses associated with collecting, storing, and distributing samples and data, but whether biobanks can request fees that go beyond mere cost recovery—and make a profit—is a relatively unexplored question in the literature, yet a pressing practical one. Would it be acceptable (although arguably not practical) that such infrastructures are set up also for the purposes of making profit—in particular, by sharing data and samples for a fee?

2. Research questions

For the present study, addressing all aspects of the commercialization controversy would have been too demanding, on account of the already broad scope of the study and of the degree of specificity and technical knowledge required to discuss IPRs and public domain in depth. Therefore, the study was designed with the secondary goal to gather respondents' reactions to broad statements about how the investigators would deal with making data public by submitting it to the public domain and simultaneously renouncing patenting rights on the data generated by the project, and on the admissibility of fees to generate profits. The proposed policy states as follows:

> The data generated by the research will be treated as a public good and all investigators must agree to put their findings and supporting data in the public domain on a regular basis and without undue delay.

We also added a statement regarding patenting rights:

> The representative of the colon cancer patients proposes the addition of a provision under which investigators agree not to exercise any rights they may have to patent a gene sequence.

Furthermore, we asked participants two questions on fees: 1) Assuming the participants were not paid, is it acceptable for the repository to ask researchers for a fee that is greater than what is needed to cover the costs? 2) If a fee beyond reimbursement is paid, should some of the additional income be shared with the participants who provide the samples and personal data?

3. Results

3.1 Is there an obligation to put the data in the public domain?

Firstly, we asked whether investigators accessing a publicly funded biobank have an obligation to put the data which they generate in the public domain. This requirement raised less controversy than expected among respondents from both samples. In fact, the majority of respondents supported the claim that researchers have an obligation to put data in the public domain. Some argued that it is an ethical obligation, because research data are an inherently public good and/or because it is a form of benefit sharing—a way to ensure that data is available to anybody, thus increasing the chances for the public to benefit from research. One respondent said:

> I think the data should be treated as a public good. Because I think the purpose of research is to advance science for the benefit of the public. I don't think the purpose of research is to advance the careers of the researchers ... [Doing good and self interest] are not mutually exclusive ... people can still advance their careers without withholding information that would benefit the public. I don't feel it is such a conflict. I also think that this is something that varies so much internationally. We have such a profit motive in [the US]. I don't know whether it is true in a lot of other countries. I think genetic research—the human genome project is an international project as supposed to benefit everybody so I don't think that individual countries and the economic motivations of individual countries should override the public good intended by genetic research (#37 IS, Oceania, social sciences, bioethics, university, A).

Other respondents were more concerned with the scientific significance of public availability of data, arguing that it is scientifically required, because science is sharing of knowledge, and/or because sharing is the most efficient way to achieve the most from the research (for instance, by avoiding duplication of studies, by validating results more quickly):

> ... the only way to move forward science to make sure it's not limited to a few investigators having access to that information but to have the full scientific community to have access to that information as long as it is not personally identified (#47 US, law, social science, government, A).

A substantial number of respondents found that the interests of researchers and the public are better served if research results are put in the public domain with a delay so that the researchers who have generated the data have exclusive use of the data for some time and they can publish them in scientific journals. Some also pointed out that putting raw data, preliminary data, and negative results in the public domain is not desirable because of the dubious scientific utility of having those data published (#80 US, genetics, bioethics, philosophy, theology, university, R/A/S: "not necessarily confirmed and it maybe misleading"; #92 US, law, medicine, genetics, university, R/A: "I am not sure that science is advanced by making every single failed step available") and/or because of academic pressure to publish (#80 US, genetics, bioethics, philosophy, theology, university, R/A/S: "competition for funds means not all researchers are getting fully funded so you cannot force them to put them [data] into the public domain that would allow a competitor to take that little piece to put it together with what he or she is doing and jump ahead"). The critics of the sharing requirement argued that (i) scientific data is private good; (ii) the requirement would remove the incentive to do research and therefore some types of research would not be carried out; and (iii) practically, implementation of the requirement would be inconsistent because private research entities would not abide by it. In sum, there was some agreement, on both sides of the argument and from respondents of both samples, that (i) the requirement to put data in the public domain is desirable, and that it ought to be imposed when research is publicly-funded, and (ii) that a delay may be appropriate so that researchers can publish the results without fear of competition.

3.2 Are patenting rights on gene sequences admissible when the research is publicly-funded?

Second, we asked whether investigators accessing a publicly funded biobank should agree not to exercise any right they may have to patent a gene sequence. A substantial number of respondents of both samples were in favor of limitations on patenting. While few opposed the patenting system altogether—based on concerns of maximization of the public good—most were opposed to patenting gene sequences and were only in favor of patents on applications or techniques or other information (#93 IS, Europe/Middle East, biology, genetics, biobank, S: "[There are] complicated implications … patenting a gene means limiting research on it, limiting pharmaceutical applications, limiting diagnosis, and many types of issues would be under monopoly in this case." In fact, to many, patents should be limited to instances in which there is "a genuine inventive step" (#82 IS, Europe, natural science, A/R). On the opposite end, other respondents argued that, in the absence of patenting rights, private companies would not invest in research and development: "…it is highly doubtful that a commercial company would want to even carry on this kind of research without intellectual property rights" (#14 IS, Oceania, political science, indigenous IP, university A/R)—and that it is the right of any investigator to patent new knowledge. Some suggested that having non-exclusive licenses issued by patent holders is a desirable arrangement because the researchers' rights are recognized without compromising the sharing of knowledge, which is granted by the requirement of full disclosure at the time the patent is filed and by licensees' free use of the invention. Finally, a handful of respondents argued that patenting arrangements should be concerned primarily with protecting participants' interests— "patients have a right to protect themselves from any possible patenting" (#93 IS, Europe/Middle East, biology, genetics, biobank, S). Participants should be involved in negotiations, and possibly hold the patent or share IPRs with the researchers. Once again, informed consent was invoked by some as the answer to this delicate balance: as long as participants are informed about, and consent to, the commercialization of the research findings, commercialization raises no issue. In sum, patenting genetic inventions is controversial and, while respondents agreed on the need to analyze the implications of patents for the public, they disagreed on which policy would better serve the public good from a practical standpoint. There was some disagreement on the extent to which participants' rights ought to be balanced against researchers' rights and the public good.

3.3 Is it admissible to collect an access fee that is greater than what is needed to cover the costs?

Overall, respondents found that allowing the repository to recover costs is acceptable. By contrast, imposing fees beyond cost recovery was controversial, and the question opened the door to a variety of far-reaching responses. In fact, in answering, respondents often broached issues ranging from the public interest in conducting scientific research, to whether scientific research can be reconciled with profitability, to the commodification of the human body, to fairness in access to scientific research opportunities. The most interesting finding is that respondents from the US sample and the international sample expressed very different views on the ethical implications of for-profit biobanks. While the majority of US sample respondents were substantially less inclined to conclude that imposing a fee beyond cost recovery raises ethical issues, several international sample respondents—and among them also three biobankers operating in Europe—were concerned that establishing commercial biobanks is ethically unacceptable. A variety of reasons were offered in support of this claim. The most commonly cited was that the intent to profit from scientific research undermines its public interest dimension. Others pointed out that publicly funded repositories or repositories conducting fundamental research have a moral obligation to share samples for free. Respondents from the international sample were also concerned that a fee-based access system would exclude from such research investigators from less affluent countries:

> … who are the rich countries that can do the research? Maybe you have one country that wants to do it but can't afford scholarships or finding or whatever to send [its people] (#37 IS, Oceania, social sciences, bioethics, university, A).

> Unacceptable because if someone offers more money for a poor question, which one will you take? The only criterion should be the quality of the scientific use of the data, not an issue of money. This should be completely prohibited (#53 IS, Europe, medicine, public health, Int. Org., R/A/S).

Moreover, while some argued that *only* private companies ought to pay a fee in order to access collected data and samples collected, no respondent suggested that the fee should be proportionate to the economic possibilities of the investigators. Finally, a few respondents (all of them part of the international sample) were concerned that a fee-based access system could result in a (unethical) form of commodification of the human body and be tantamount to a trade in body parts.

Supporters of a fee-based access system argued that such a system in itself does not harm participants, and thus it does not raise particular ethical issues:

> I am a participant and I volunteer to give my sample so I don't care about what will happen at the other level. The only important thing to me is that my anonymity is protected and that beneficial results come out, I will be contacted to know this. Whether the organization

makes certain profits out of the repository, that is up to the organization itself and the controlling agencies of the country where it is hosted (#93 IS, Europe/Middle East, biology, genetics, biobank, S).

Indeed, some pointed out that imposing a fee could benefit participants if the fee is shared with individual participants as a personal benefit, or with the group that participated in research as a form of collective benefit, or reinvested in research. However, one respondent offered an interesting argument against sharing the fee with the participants, reasoning that there should be no specific financial advantage in being chosen as an individual participant:

> ... because the participants individually are recruited to the repository as representatives of a class of people, and it's a matter of chance that it is you rather than me or me rather than you that gets recruited—it depends on where I live, whether I consulted a physician that is interested in what you have—there should not be a direct financial return or material return to the participant in that way. It might be possible, if there is significant commercial gain to be had, to devise some form of benefit sharing arrangement that benefit the class of people but not the individual (#82 IS, Europe, natural science, A/R).

Finally, a substantial number of respondents from both samples agreed that the informed consent form must indicate that the biobank could make a profit and that a proper informed consent process makes this model less ethically questionable.

4. Analysis

The implications of the commercialization of genetic research are numerous and controversial. In this study, we were able to address only three relevant questions: the ethical obligation to put genetic data in the public domain; the admissibility of patenting rights when the research is publicly-funded; and the admissibility of access fees that are greater than what is needed to cover the operating costs of a repository. While respondents supported a requirement to put the results of research on genetic data in the public domain after a reasonable time that allows researchers exclusive access to results and an opportunity to analyze and publish them, patents and fees were very controversial. Responses were informed by respondents' divergent views on the proper goals of scientific research and the proper conditions under which research ought to be conducted. Respondents for less developed countries were concerned that commercialization may impose barriers to access essential research for researchers of countries that cannot afford to file for patents or to pay access fees. More broadly, respondents also presented contrasting views of the role that research ought to play in society. To some, scientific research is undermined if sustained by the profit motive. Scientific research is meant to advance knowledge in the public interest, and commercial interest undermines its basic nature. By contrast, others noted that having commercial interests *per se* is ethically neutral and often provides the necessary incentives. Lacking these, research would move at a slower pace or, for some health problems, would not be carried out at all.

5. Implications and conclusions

The findings of this study suggest that commercialization is a very controversial topic. The different views are difficult to reconcile because they reflect substantial disagreements about the basic purpose of scientific research and the appropriate conditions for its conduct. Therefore, the debate on patenting rights and commercial exploitation of genetic research is not easily settled. Indeed controversies may well heat up in the future, given the increasing need for studies involving multiple populations in different regions of the world.

How to deal with disagreement and controversy? Firstly, more research is certainly needed to uncover the roots of these disagreements and their implications for policy. Secondly, it is important to realize that widely accepted ethical principles—in particular the principles of respect for autonomy and justice—are conceptual tools that should play an important role in the analysis of some of the most burning issues relating to commercialization. Respondents pointed out that the duty to establish a relationship with research participants based on a candid disclosure of commercial interests and other implications of a study for which participants are recruited, is a necessary element of an ethically defensible approach to biobanking. Moreover, concerns that commercialization may be an obstacle to a fair access to research and a fair distribution of the benefits of research have emerged from the interviews, especially among respondents from less developed countries. It follows that considerations of justice cannot be overlooked in the commercialization debate. Thirdly, if disagreements cannot be ultimately reconciled, opinion leaders in biobanking and genetic research must be at least aware of the diversity of views and of what is unacceptable to individuals with opposing views.

Bibliography

Barton, J.H. (2007), *New Trends in Technology Transfer and Their Implications for National and International Policy.* International Centre for Trade and Sustainable Development (ICTSD).

Burton B. (2002), "Opposition Stalls Genetic Profiling Plan for Tonga," Inter Press Service, 18 February. Available at <http://www.commondreams.org/headlines02/0218-02.htm>

European Parliament and Council (1998), Directive 98/44/EC on the Legal Protection of Biotechnological Interventions.

Foray, D. (2004), *The Economics of Knowledge* (Boston, MA: The MIT Press).

Hopkins, M.M., Mahdi, S., Patel, P. and Thomas, S.M. (2007), "DNA patenting: the end of an era?," *Nat Biotech* 25, 185-87.

Jensen, K. and Murray, F. (2005), "Enhanced: intellectual property landscape of the human genome," *Science* 310, 239-40.

Lowrance, W.W. and Collins, F.S. (2007), "Identifiability in genomic research," *Science* 317, 600-602s.

Marth, G. et al. (2001), "Single-nucleotide polymorphisms in the public domain: how useful are they?," *Nat Genet* 27, 371-72.

Rai, A. and Boyle, J. (2007), "Synthetic biology: caught between property rights, the public domain, and the commons," *PLoS Biology* 5, e58.

Riordan, T. (1995), "A recent patent on a Papua New Guinea tribe's cell line prompts outrage and charges of 'biopiracy,'" *The New York Times*, 27 November.

South and Meso American Indian Rights Center (1995), Declaration of Indigenous Peoples of the Western Hemisphere Regarding the Human Genome Diversity Project Available at <http://www.indians.org/welker/genome.htm>.

Watts, G. (2007), "The locked code," *BMJ* 334, 1032-33.

Chapter 13

Benefit-sharing and Remuneration

Agomoni Ganguli-Mitra

1. Introduction

1.1 Why share benefits?

The notion of benefit-sharing is certainly not new to research ethics. In the field of genetic research, however, the concept has gained prominence through different channels. One of the earliest efforts to establish benefit-sharing in genetics bore fruit at the 1992 Convention on Biodiversity (CBD) at the Rio Earth Summit. The other, almost parallel emergence of benefit-sharing as an important aspect of genetic research was witnessed with the development of the Human Genome Project and a statement by the Human Genome Organisation (HUGO 2000). While, the CBD was specifically concerned about plant and agricultural genomics and as such relied on the notion of ownership of biological material, the Human Genome Project and the HUGO, however, adhered to the notion of the human genome as a "common heritage of humanity" and stressed that as a result all of humanity should benefit from its exploitation.

However, benefits to humanity does not necessarily imply that individuals or communities who have participated in particular research projects benefit from it, nor that the progress made by research will reach those who need it most. Although participating in research is still largely considered altruistic (Berg 2001, 242), there have been calls in recent years to address global inequity in biomedical research, and it is largely recognized that the benefits of biomedical research should be used to compensate those less well-off (Simm 2007, 164), especially where the private sector and industrialized countries stand to make profit from the contributions of individuals from developing countries (Dickenson 2004, 109, Schulz-Baldes et al. 2007, 8). The benefit-sharing discourse has trickled down from traditional biomedical research into genomic research too, since genomics is a potentially highly lucrative market and since such research also tends to seek out isolated populations and groups in developing countries, who often suffer from acute health and economic inequity.

A related concept is the idea of benefit-sharing based on reciprocity. The principle of reciprocity, widely used in traditional biomedical research, implies that participants should be compensated in some ways for going through the inconvenience of research and for taking risks. However, the concepts of inconvenience and risk tend to lose some of their strength in the context of genetic research (where inconvenience and risk of physical harm are minimal). Moreover, as one author points out "the risk discourse delineates a recipient community and those responsible for creating these risks have a duty to compensate within the reciprocal setting" (Simm 2007, 162). In genomic research, both the concept of risk and the delineation of the participating group can be tricky.

Among current international guidelines, not all address the topic of benefit-sharing in depth. It is worth noting, as Knoppers and Abdul-Rahman do in chapter 2 that the HUGO, although adhering to the notion of a common heritage, does provide for the recognition for altruistic participation through various forms of benefit-sharing (see chapter 2, section 2). As noted by Vayena et al., in chapter 3, UNESCO's International Declaration on Human Genetic Data recommends sharing benefits (Art. 19) for society as a whole and the international community. Other guidelines also touch upon the subject (see Chapter 3, sect. 5), however most international guidelines lack explicit and harmonized guidance on this matter.

1.2 Procedural difficulties surrounding benefit-sharing

Be it based on one or a combination of the concepts discussed above, benefit-sharing is emerging as a key topic in genomic research, and is therefore of high relevance to biobanks. What remains extremely difficult to navigate through, however, are the specific procedural and practical aspects of benefit-sharing. Some of the main problems are summarized below as example questions, grouped under the following four related yet distinct topics: when, who, what and how?

(a) When? It is not clear when exactly the obligation to share benefits arises. For example: should benefit-sharing always take place between sponsors/researchers and participants? Or only when the private sector is involved? When there is an economic disparity between the two parties or more specifically only if profits have been made from a specific research project? Should benefit-sharing take place only where samples have been collected for a specific research project with clear goals or also in general cases, for example where samples are collected (as add-ons) during an unrelated clinical trial (Pullman and Latus 2003, 242-3), or constitute a study on previously archived material?

(b) Who? Who should be the recipients of a benefit-sharing scheme? The individuals who participated? What, if anything is owed to those living in the same community (or are part of the same patient-group) but did not donate samples? Should certain people profit from the fact that they have a certain genetic make-up? Should benefits then rather target humanity as a whole?

(c) What? What constitutes a fair benefit? A percentage of profits from the research, rights to its intellectual property, tests, medication or treatment made available from the research, capacity building or attention to the basic (health) needs of a population?

(d) How? How to make sure benefits are shared without making them undue inducement? How to identify the individuals and institutions through whom we might negotiate the details of benefit-sharing and through whom we might share benefits with the population?

1.3 How the topic was addressed in the study

In our project, the topic of benefit-sharing was explored in the same scenario as collective consent (see Chapter 7). Researchers wishing to collect samples for commercial pharmacogenetics studies approach an indigenous population. The size and location of this population are unspecified, other than being located in the same country as the researchers. Having identified the group's representatives in the form of a governing council, the researchers first approach them as gatekeepers to the community. The following schemes are offered to the community as possible benefit-sharing options:

Option A: Making any genetic tests resulting from the research available for free to the indigenous group for ten years.

Option B: Making an annual donation for a period of ten years to the hospital that provides health care to the indigenous group, of a sum equivalent to 3% of the revenues generated by any intellectual property rights resulting from the research.

Option C: Donating several pieces of durable medical equipment to the hospital that provides health care to the indigenous group.

The governing council rules out all these options and determines that it should rather own all intellectual property rights arising from this research (*Option D*). The negotiations fail and the governing council refuses, on behalf of the entire population, to participate in research. The researchers then consider whether to bypass the governing council and approach individual members for participation in exchange for US $800 (a month's salary) as remuneration. Respondents' opinions were further probed with additional questions such as whether their answers where based on practical considerations or on ethical principles and whether they saw a significant distinction between benefit-sharing and fixed compensation for participation.

Although the core of the debate regarding benefit-sharing and remuneration revolved around this scenario (scenario C), the topic was also touched upon in another question (scenario B), based on a population database. The question asked was if the database made profits from the circulation of genetic material, whether that income then should be shared with all participants.

2. Results

2.1 General observations regarding benefit-sharing

A first glance at the interview results on the topic of benefit-sharing shows that there is very little consensus among respondents regarding which are the best procedures to follow for benefit-sharing. It must be noted however, that although strong opposition was expressed against certain types of benefit-sharing and remuneration, there was little opposition against the notion of sharing benefits that accrue from research. One exception can be cited here:

> In my country people are against any payment in exchange for contributing to the advancement of science and people are not interested in these payments ... People's benefit is advancing knowledge (#48 IS, No. Am./Europe, philosophy, university, R/A).

A remark may be due here that the respondent is sharing experiences from a Western industrialized country and is not necessarily thinking in the context of less well-off indigenous populations. Later, he discusses the various options of benefit-sharing and chooses the durable medical equipments (c), which he believes would be the most useful contribution to the community. Almost all the other respondents, however, agreed that benefits should be shared:

> We have to continue to make sure that resources are available to people who are contributing to the development of infrastructure or profit for whatever company or university or institution or government. That is the basic principle here. It is the principle of sharing (#12 US, Africa, genetics, university, R/A/S).

A small percentage from this group encouraged stricter measures as to when benefits should be shared, such as only when profit is made or when the private sector is involved and stands to make large profits:

> What is important is that private companies share a percentage or a fixed amount with the society that has made the samples available. Whether a percentage or a fixed amount, I think it depends on the practical case and on the benefits that people expect. However, what is important and indispensable is that private companies share the benefits (#31 IS, Europe, medicine, university, S/A).

2.2 The scope of benefit-sharing

One reason why little pattern or preference could be detected in the responses, was the fact that the choice of a suitable benefit-sharing scheme was also seen as dependent on factors beyond the nature of the study. One such factor is the context in which the study is being conducted. Context-specificity was well expressed in the words of one respondent:

> Biotech industry in [my country] is not likely to be substantially formed by members of the indigenous population. In contrast, in Switzerland, if one canton or whatever has one biotech company incorporated, it is not unlikely that a significant number of the residents are shareholders of that company. In the United States, if you consider the American Indians, they may not be the shareholders. So how you then negotiate benefits will depend on the relationship between the indigenous population and the owner of the biotech company. If you take it from the very crude situation of a very poor indigenous population that has a high-tech biotech company established in their midst, you may want to say that that company should be socially responsible ... (#81 IS, No. Am./Africa, medicine, bioethics, university, S).

It is perhaps worth noting that only a minority among respondents considered the different economic levels among various indigenous populations. Perhaps due to the wording of this scenario, most respondents modeled their answers around economically and educationally less developed populations, often citing the examples of tribal communities.

Another point that had particular resonance among interviewees was the scope of benefits. A great majority agreed that benefit-sharing should essentially target communities rather than individuals, a position that was later supported by the reluctance of respondents to endorse individual remuneration. As with collective consent (see chapter 6), it was also suggested that the nature of the study itself defined the nature of benefit-sharing:

> If you go taking samples from an indigenous population, your reason to do it is inherently value-based ... you have chosen the community as opposed to taking 200,000 samples from people who happen to specifically live somewhere ... If it is collective research, the benefits therefore need to be collective (#14 IS, Oceania, political science, indigenous IP, university A/R).

Moreover, some felt that if such a scheme is to be both realistic and useful, it is perhaps best to leave the choice to the community as to the most appropriate scheme for it. The point of view of outsiders, such as sponsors, researchers and in this case the interviewees themselves may not be appropriate:

> ... each of these are options, in my mind, is coming from Biotech worldview of how we benefit the donor community and my basic position is that if you go into a group, you first have to know the group and find out, and let them decide what would make for fair ... (#15 US, genetics, university, R/A/S).

However, some respondents from this group further emphasized that a fair negotiation of benefits would require the involvement of the entire group and not just discussions with the governing council (#46 IS, No. Am., law, social science, NGO, A).

As expected from the questions raised in the literature, the discussion regarding the scope of benefit-sharing also touched upon the problem of group delineation and the difficulty of deciding who is entitled to the benefits of a particular research study.

> But simply getting the samples from the 2000 people does not seem to me to obligate you to provide free tests for the other 3000 members of the population in that location because there is nothing special about their being in that location, unless there was some participatory role. It is like saying if 5000 people pre-disposed to cystic fibrosis gave their samples we should provide in the US free genetic test for everyone pre-disposed but you have to have a justifying reason for that (#55 US, philosophy, bioethics, university, N).

2.3 General dissatisfaction with benefit-sharing schemes

About 20% of the respondents specifically expressed dissatisfaction with the benefit-sharing options offered in this scenario, although all of them eventually settled for one or more of the schemes presented. Some preferred modified versions of the options, such as lifting the ten year time limit on options A and B or suggesting that several of the options be offered together. It was also not uncommon for such respondents to point out the exploitative nature of the arrangements presented here, echoing calls in the literature to address issues of justice through benefit-sharing:

> I think they [the options] are band-aids ... You don't fix a problem, you just put something on it and hope it goes away ... This whole question reeks of prejudice and power differentials, and it worries me a lot because I think people consider the indigenous

populations less educated and less able to defend themselves in sort of a westernized world. It really bothers me they'll offer them a little something and say: 'ok now we have done our part and we can go and make millions off this data' (#64 US, medicine, bioethics, university hospital, N).

Beyond the general comments regarding benefit-sharing, it was also interesting to observe the extreme variation in the responses to the specific options presented. It is worth pointing out that none of the choices attracted a majority of respondents and reasons supporting the choices were often more varied than the range of preferences themselves. The following is a summary of the opinions expressed for each option.

2.4 Option A: offering free genetic tests for ten years

Often, reasons for favoring option A had to do with the fact that it was closely related to the aims of the project, or in the words of one respondent: "because it has to do with research itself" (#99 US, medicine, genetics, university hospital, R/S). More so, perhaps in light of the fact that many of the other options were related to sharing financial profits:

> … it is not a natural connection that people with a particular disease ought to profit from that disease. What they need is to get the test for that disease and whatever treatments (#55 US, philosophy, bioethics, university, N).

Many respondents in this group seemed to base their preference on the principle of reciprocity, echoing the existing guidelines and principles on research ethics in general. Participants would be receiving a benefit related to what they contributed towards, almost as a direct "feedback" from the research (#13(c) IS, No. Am. /Asia, law, university, A/R). Of interest were also reasons which related option A to a direct benefit for the individual, rather than depending on an indirect source such as a representative body: "I prefer the individual benefit options to hoping that the governing council's working in their interests" (#18 US, genetics, NGO, R/A/S)

However, many more respondents were against option A. In direct opposition to the argument presented above, one respondent gave their reason for rejecting A as follows:

> The general view is that people make a mistake in trying to link the benefits to a particular thing such as the nature of the study; it might be terribly inefficient to do so, it might be that the people could use the money to [do] other things that would be of greater use to them than just the genetic tests and that might be better for everybody (#25 US, philosophy, government, N).

Others felt that option A was paltry (#35 IS, Europe/So. Am., medicine, genetics, university, S), especially in face of the potential for profits from this research. It is interesting that those who followed the reciprocity principle, as above, seemed to find option A adequate where as those who tended to argue along the lines of fairness and equity seem to think of it as a meager compensation.

Concerns were raised about the contingent nature of option A, or that A seemed like a "bet" (#56 IS, Europe/Europe and Africa, medicine, life sciences, government, hospital, S/A/R), a criticism also leveled at option B (sharing a profit of 3%) as many felt that such a contingent option would not be beneficial or fair to the population. Others, it must be noted, felt that it was precisely the contingent nature of the scheme that made it benefit-sharing, as opposed to payment for research.

Finally, some felt that genetic tests would not be useful to this population (#61 IS, Africa and Europe/Africa, life sciences, university, R/U/A/O), or even if such a test is offered, it might be for a disease that has no cure or for which treatment is too expensive (#33 US; bioethics, medicine, government, A). Comparatively, an indigenous population such as the one described here would be better off with more basic health benefits (#09 US, philosophy, catholic theology, university, N)

2.5 Option B: sharing 3% of profits with the local hospital

Option B tended to come up more often as the preferred choice, or at least as the most acceptable among all the options given. The main reason for favoring option B was also related to notions of fairness. It seemed right to many respondents that disadvantaged groups should not continue to suffer from inequities, especially if they have made a significant contribution:

> I always feel that there has been a lot of abuse of certain groups of people that had either peculiar mutations or genetic polymorphism, especially by companies. The companies did not take care to do something in return to them. In the Western civilization, [donors] should not get a benefit out of it (#30 IS, Europe, medicine, bioethics, government, R/A/S).

Moreover, option B represented a long-term relationship between researchers and the population and as such illustrated well the spirit of benefit-sharing:

> I think B is the most acceptable because this involves some continuing relationship with the indigenous group, and recognition of responsibilities to share benefits (#02 US, philosophy, catholic theology, university, N).

Many of the proponents of option B however, noted that they were not certain whether 3% was an appropriate amount in this case (#06; #19; #25; #28; #80). Some of the criticisms against option B were based on its contingent nature, as noted earlier, but more importantly its monetary, and potentially exploitative nature (see also 2.9):

> I don't strongly endorse the HUGO view that there should be direct profit sharing with the people who participate. I think that can lead to some fairly negative things ... I think getting over to a larger sum [is] to pay for access, which means that you are not necessarily helping them but exploiting them by paying them (#68 IS, No. Am./Europe (nationality: No. Am.), life sciences, consultancy, R/A).

2.6 Option C: donation of durable medical equipment to the hospital

Those who opted for option C perceived this scheme as most beneficial to the group (#73 IS, No. Am./Asia, life science, government, S) and sometimes also as benefit-sharing in the form of capacity-building:

Ideally what you would like to see is something that looks like capacity building as well as something that contributes to the health of the people.(#92 US, law, medicine, genetics, university, R/A).

Also, in line with those who were against the contingent nature of some of the proposed schemes, many saw this as a more reliable and therefore better choice:

Well, certainly if I was advising the Governing council looking after its interests … It seems to me that the one that offers the indigenous people the most likelihood of actually obtaining something for this activity is option c … I am not sure if that itself is enough but you could imagine something more substantial … as you know most research doesn't result in anything. And so they're going to wind up with nothing most likely (#59 US, bioethics, protestant theology, philosophy, university, R/A).

Interestingly, this option was also rejected precisely because it meant: "something now and nothing in the future" (#74 IS, Asia, genetic, hospital, S). Other reasons for rejecting option C were that it was seen as not enough, "inadequate" (#12 US, Africa, genetics, university, R/A/S) or even that it "is tokenism. It is insulting" (#01 US, medicine, bioethics, university, N). Similarly to those who preferred free genetic tests because they represented a direct benefit to those who participated, some interviewees ruled out C because it did not:

… a donation to the hospital may or may not reach the participants. I want a direct benefit to the participants (#93 IS, Europe/Middle East, biology, genetics, biobank, S).

The least suitable is C because there is no defined relationship between the results generated and the benefit (#17 IS, Europe, law, university, R/A).

Finally, to some, the lack of logical relationship between the research and the benefit, as well as the latter's fixed nature (see also 2.8), meant that option C almost felt like a "kind of bribe" (#50 US, catholic theology, genetics, philosophy, bioethics, university, A), emphasizing, as others did before that certain types of benefits or certain amounts of benefit may lead to an exploitative relationship rather than an acknowledgement for participation or a fair procedure for sharing the benefits of a study.

2.7 Option D: ownership of IP rights

As expected, option D drew a wide range of opinions and illustrated some of the more difficult aspects of regulating genomic research and databases. Whereas a few people found D to be the best option compared to others, which were "too stingy" (#54 US, bioethics, protestant theology, philosophy, university, R/A), many more people had trouble accepting D as it had been presented. Some felt that option D was the best choice in principle, but utopian (#35 IS, Europe/So. Am., medicine, genetics, university, S), or that given the involvement of a private company, it may act as disincentive (#72 IS Europe/Africa and Europe, (nationality: Europe), medicine, S/A).

The US interviewees were also asked whether they then preferred option D as shared IP rights between Biotech and the governing council. This many of them preferred, sometimes even putting option D (modified) on top of their preference list.

Others, from both sets of respondents ruled out D as a matter of principle, as they felt that IP belonged rightfully to the funding body (#22 IS, Africa, science, self-employed, A; #10 US, humanities, law, university, R/A) or according to some, to the intellectual process (#61 IS, Africa, and Europe/Africa, life sciences, university, R/U/A/O, #94 IS, Asia, life sciences, government, A). Donation of samples, according to these respondents did not qualify for IP rights:

> The people that are giving the samples are not really contributing; they are just giving the samples. They are entitled to benefit-sharing but not to [IP rights] because they didn't do any work. They can get a sum of money [as benefit-sharing] but not the patent (#74 IS, Asia, genetics, hospital, S).

D also brought up one of the more delicate topics in genomic research, that of the ownership of biological material. This issue cut across several scenarios and here too, many respondents pointed out that option D was "qualitatively different because it transfers ownership" (#38 US, bioethics, philosophy, theology, medicine, R/A). Opinions ranged from the specific, that "there ought not to be private ownership of an indigenous group's biological information" (#38 US, bioethics, philosophy, theology, medicine, R/A) to the more general view that human genetic material should not be patented:

> One of the topics from the whole interview I found the most challenging is this concept of sharing IP rights or profits with participants. I think it raises a very difficult question: to what degree do we own our bodies and our tissues and our samples as a resource. And I think it could be a slippery slope towards us perceiving individuals as walking-talking databanks of genetic information (#46 IS, No. Am., law, social science, NGO, A).

2.8 Various approaches to benefit-sharing

Clearly, whether respondents were inclined to choose one option rather than another, points of view ranged from the practical to the principled. On one side were those who felt that adequate benefit-sharing schemes depended entirely on practical or economic considerations:

> ... the circumstance is a lot like one of investing. You could really go wrong with an investment if you take risks, which is the way I see the sharing part here. Or you could get immediate gains if you take much less risk, which would be ... fixed compensation. So I think it depends on your level of risk tolerance (#95 US, bioethics, philosophy, theology, university, N).

> I am not sure I would do this as an ethicist. I think I would do this as a financial person. I would imagine an economist would tell these people don't do all this future stuff because you're basically not going to get anything. It's not as much an ethical issue as it's really bad financial stuff (#59, US, medicine, bioethics, government, R/A/S).

On the other side were those who were particularly concerned that benefit-sharing should not become a business-like negotiation:

> The difference should be relevant in the sense I am going for sharing not in the sense of business, sharing here as proposed of research findings and how that could be put into practice not in the business sense of making money, but in the sense of helping them out, the whole community as such not material exchange or cash exchange (#05 US, Asia, philosophy, bioethics, theology university, N).

It is also interesting to see that the concept of benefit-sharing differs significantly from one person to the other. When asked about the distinction between fixed compensation and benefit-sharing, many felt that benefit-sharing was morally superior to fixed compensation. However, the definition of the two terms, as well as what counts as fixed rather than benefit-sharing differed according to the respondent. Some felt that only options that were contingent (such as B) or directly related to research (such as A) counted as benefit-sharing (#03 IS, So. Am., nat. sciences, genetics, bioethics, university, A; #46 IS, No. Am., law, social science, NGO, A). Others felt that only options which benefited the community rather than individuals counted as benefit-sharing (#13; #19; #24; #73), so that, for example, benefiting from research does not depend on the "luck" of having the gene or not (#66 IS, Europe, law, consultancy, A/R). Some felt that benefit-sharing required elements of a long-term relationship between researchers and the community (#19 IS, Europe/Asia, philosophy, university, A; #26 US, medicine, social sciences, government, R/A) and others that it should be essentially knowledge-based rather than on financial contributions (#63 US, bioethics, medicine, catholic theology, university, A)

Related to this was the concern that financial compensation, if too large, might lead to payment for research and to exploitation and might therefore be morally unacceptable (#68 IS, No. Am./Europe (nationality: No. Am.), life sciences, consultancy, R/A), such as in the case of option B. Others felt that it was precisely because the amount was not fixed that made option B a "proportional sharing of the benefit" (#45 US, bioethics, philosophy, government, O). Exceptionally, one respondent felt that only option D (modified) was benefit-sharing, because it represented co-ownership (#64 US, medicine, bioethics, university hospital, N).

2.9 Benefit-sharing, remuneration and exploitation

A recurrent theme in the exchanges on benefit-sharing was exploitation. While many saw benefit-sharing as an important instrument to mitigate the exploitative nature of some research arrangements, concerns were also raised that benefit-sharing should not itself become an instrument of exploitation. As with collective consent (see Chapter 7) respondents warned against giving too much power in the hands of a representative body (#19 IS, Europe/Middle East, genetics, university, S), while recognizing the delicate nature of such a negotiation and its potential to become a form of undue inducement, or never reach those who are entitled to it:

> How do we know that this group is representative of the people who are participating in the project? On the other hand, I am also nervous about negotiating directly with

individual participants because often they got cheated, they got exploited by these research teams or these biotech corporations or whatever … How to make it in a way that there is fairness and justice and is given proper share of benefits to those who are participating in such research projects, I think it is a big ethical issue … There is an ethical obligation of whoever is doing this to make sure that this is fair to the individual and not saying 'you know the money, I don't care at the end who gets it' (#21 IS, Europe/Middle East, genetics, university, S).

Related to this was the uneasiness concerning exploitation: in other words that certain forms or procedures of benefit-sharing may in fact harm participants. This was one reason some preferred indirect (#62 IS, Europe and No. Am./Europe, medicine, catholic theology, university, A), or non-monetary benefits (#65 So. Am., life sciences, genetics, university, A) and that some respondents specifically warned against benefit-sharing becoming payment or inducement (#06 IS, So. Am., philosophy, bioethics, university, A; #68 IS, No. Am/Europe (nationality: No. Am.), life sciences, consultancy, R/A). This perhaps also explains that perhaps the only issue that saw a majority of interviewees reply along similar lines was the questions on remuneration, that is, when asked whether offering an individual remuneration of $800 would be acceptable. While a small minority took a weaker stance against remuneration:

It should be permissible because it is a judgment for the individual. I know that indigenous people are not really advanced in terms of [not understandable]. I don't like it but it should be permissible (#94 IS, Asia, life sciences, government, A).

A much larger number replied with a categorical "no" to this option (#03; #13; #14; #21 …). Similar results can be observed for the related question in scenario B, where participants were asked whether it would be acceptable for a biobank to make profits if the latter were shared with participants. Once again a clear majority said no, except those who associated this type of remuneration with some forms of benefit-sharing.

2.10 Results from the WHO consultation

During the WHO consultation (see Chapter 6), the topic of benefit-sharing was directly and indirectly discussed in various places (see chapter 6, sect. 2.3 and 2.6). Related to the right to intellectual property, some experts felt that, although an important incentive for research, the IP often put the developing countries at a disadvantage. Beyond the various options offered in the scenario, it was pointed out that certain forms of feedback to participants, working in collaboration with the local community, fostering local research and capacity building as well as offering general medical services could also be suitable forms of benefit-sharing. Echoing the concerns expressed by the interviewees, experts pointed out that there was always a risk of undue inducement through benefit-sharing, which might compromise consent.

3. Discussion and conclusions

Perhaps one of the most important remarks that can be made regarding the results on benefit-sharing is that it is almost impossible to find consistently recurring patterns between the choices and the reason given for them. While almost all respondents agreed that some form of benefit-sharing scheme should exist, there was little consensus regarding the questions posed in the introduction: when, who, what and how?

While some interviewees based their argumentation on concepts of reciprocity and benefits according to contribution, others based their thoughts on more general principles of fairness and justice. Others still preferred to follow entirely practical approaches. Calls for leaving the choice of benefit-sharing to the population or group in question were balanced by concerns that some processes of benefit-sharing might lead to exploitation, both external (coming from sponsors or researchers) and internal (from the governing council). These lines of argumentation came across while respondents discussed the merits of each benefit-sharing option, but interestingly, they often lead to different choices and preferences. While access to free genetic tests was seen by some as a fair compensation, directly related to research, it was also ruled out as meager in the face of the financial potential of genomic research. Sharing a percentage of the profits that accrue from research was therefore often seen as substantial as well as more flexible for the population, but it also raised concerns of exploitation and a slippery slope into financial incentives for participation in research. Medical equipment was seen as both directly contributing to the welfare of the community but also as paltry and unlikely to reach those who contributed to the research. Finally ownership of intellectual property rights was one of the most discussed choices, raising questions of ownership of biological material, disincentives for researchers as well as worries of internal and external exploitation.

Given the often raised concern regarding undue inducement and exploitation in both the interview results and the expert consultation, benefit-sharing comes across as a very delicate procedure to navigate through. While it is difficult to draw clear propositions for guidelines or an international framework, it is clear that the need for benefit-sharing is widely supported and that its procedural intricacies require further in-depth global discussion.

Bibliography

Berg, K. (2001), "The ethics of benefit-sharing," *Clin. Genet.* 59, 240-3.

CBD (1992), Convention on Biological Diversity, 5 June, Rio de Janeiro <http://www.cbd.int/convention/convention.shtml>

Dickenson, D.(2004), "Consent, commodification and benefit-sharing in genetic research," *Developing World Bioethics* 4: 2, 109-23.

HUGO [Human Genome Organisation] Ethics Committee (1998), "Statement on DNA Sampling: Control and Access" <http://www.hugo-international.org/Statement_on_DNA_Sampling.htm>, accessed 20 September 2007.

HUGO [Human Genome Organisation] Ethics Committee (2000), "Statement on Benefit-Sharing" <http://www.hugo-international.org/Statement_on_Benefit_ Sharing.htm>, accessed 20 September 2007.

Pullman, D. and Latus, A. (2003), "Clinical trials, genetic add-ons, and the question of benefit-sharing," *Lancet* 362, 242-44.

Schulz-Baldes A, Vayena E. and Biller-Andorno N. (2007), "Sharing benefits in international health research: research capacity building as an example of an indirect collective benefit," *EMBO Reports*, 8:1, 8-13.

Simm, K. (2007), "Benefit-Sharing and Biobanks," in Matti Häyry et al. (eds), *The Ethics and Governance of Human Genetic Databases: European Perspectives* (Cambridge: Cambridge University Press), pp. 159-69.

Wilson, S. and Chadwick, R. (2007), "Pursuing equality: questions of social justice and populations genomics," in Matti Häyry et al. (eds), *The Ethics and Governance of Human Genetic Databases: European Perspectives* (Cambridge: Cambridge University Press), pp. 150-58.

Chapter 14

Transfer of Samples and Sharing of Results: Requirements Imposed on Researchers

Andrea Boggio

1. Requirements imposed on researchers: transfer of samples and sharing of results

Biobanks and biological repositories are primarily research infrastructures that store samples (possibly along with data) for the use of investigators not necessarily affiliated with the biobank or repository. Samples can be made available to non-affiliated investigators by using a variety of arrangements. The people who set up a biobank must make a fundamental choice whether samples will be analyzed at the biobank on demand or shipped to the laboratory of the non-affiliated investigator. The second option raises several issues: the recipient may be tempted to transfer the samples to third parties or to use the sample for purposes other than the one mentioned at the time the request was submitted to the biobank, either of which might breach the obligation to protect sample sources from uses not specified in the informed consent. Alternatively, the recipient might decide to store the sample along with other samples and build up a parallel biobank which could undermine respect for the efforts and investment made by the creators of the original biobank to collect and store the samples appropriately. Finally, human tissue is often seen as a precious resource, which is difficult to collect and to store: misusing or overusing samples raises an issue of optimal allocation of resources. Do researchers have an obligation to handle samples in responsible ways?

Even if samples are not physically transferred outside the biobank, researchers who access samples are in the position to enrich the biobank and facilitate future research by providing data (such as genetic analyses of the samples) to the organized collection of samples and data maintained by the biobank once they have completed their research on the samples from the biobank. Do researchers have an obligation to return this data to the biobank?

The present study addresses some of these questions. In particular, the research instruments focused on (1) the investigators' obligation not to transfer samples received from a biobank to third parties without consent of the biobank and (2) the investigators' obligation to share with the biobank data and/or findings derived from the biobank samples. To explore respondents' opinions on these questions, Scenario A stated that the steering committee of a biobank proposes including the following provisions in its Material Transfer Agreement, or MTA (a contract setting the terms under which an investigator is given access to biological specimens and the associated information):

> (a) Investigators must not transfer the DNA and the associated information to persons not named in the Material Transfer Agreement
>
> (b) Investigators are under the obligation to share all research findings and the data produced for each sample with the repository

The respondents were asked whether either of the proposed policies were necessary or desirable, and they were asked whether the first policy could be enforced. In the two sections that follow, we present their responses and discuss the justifications and practical implications associated with these two provisions, namely, not transferring samples further and having to share the research findings with the biobanks that supplied the samples.

2. Results: transfer of samples to investigators outside the biobank

Biobanks are important tools for research in a variety of biomedical fields. For biobanks to be fully exploited, scientists should arguably be in a position to have broad access to the genetic materials and data that are stored in the various biobanks. Therefore, commentators have argued that, as existing projects demonstrate, the transfer of the samples stored in a biobank to investigators outside the biobank seems an inevitable step to generate scientific knowledge (Rotimi et al. 2007, Merz et al. 2002). The distribution of physical resources to the scientific community increases the likelihood that the biological samples collected will lead to valuable scientific findings. Although the view that biomedical research is a public good is widely shared, its practical implications are controversial. In this section, we focus on the question of whether sharing biological samples creates any obligation on the receiving scientists and in particular whether the preference for sharing can reasonably be encumbered by an obligation not to transfer the material further without permission. If such obligation arises, imposing contractual restrictions on recipients seems an attractive arrangement to legally uphold it. Through an MTA, a researcher who wants to receive material for his or her own studies can do so provided the researcher (and/or his or her institution) agrees to be bound by certain conditions as to what may be done with the material and what duties it has to the organization providing the material (Rodriguez 2005).

Respondents from both the international sample and the US sample substantially agreed that investigators receiving biological material from a biobank must not transfer it to institutions or researchers that are not parties to, or named in, the MTA. The reasons offered in support of this policy included the prevention of abuses, protecting participants' autonomy (#46 IS, No. Am., law, social science, NGO, A: "there is an incredible amount of secondary uses that are not authorized particularly in the informed consent"), and the fact that the donor entrusted the repository with the sample to make good use of it (#37 IS, Oceania, social sciences, bioethics, university,

A: "when you are giving DNA … you think is only going to go to that group that you have given [the sample] to"). In particular, some respondents reasoned that, if control mechanisms are not present, samples could fall into the "wrong hands." Moreover, this responsibility is perceived as being obligatory on those to whom the donors have entrusted the stored genetic material. One biobanker offered the following explanation:

> We call it policing tissue. Because the tissue bank is custodian of the tissue and it received the permission from the donor and the next-to-kin to use the tissue, the tissue bank is also the person who is responsible and has the knowledge of who is using the tissue … I have heard about several cases of tissue banks that gave tissue for research and the tissue ended up in the pharmaceutical company being used for something completely different … we have the responsibility that the tissue is used according to agreement between the bank and the donor (#30 IS, Europe, medicine, bioethics, government, R/A/S).

Most of the respondents' answers make clear—or reasonably imply—that scientists ought to sign an MTA or similar agreement that restricts their passing on to others the samples they receive for research purposes from a biobank. Respondents' answers were also influenced by a biobank being publicly funded.

Only a handful of respondents from both samples disagreed with this policy. One pointed out that in terms of research it is often difficult to establish all future collaboration right at the outset, and that a narrowly drafted MTA could result in an excessive burden for researchers. To redress this inconvenience, other respondents suggested that researchers interested in sharing the material with researchers not named in the MTA may simply request that the biobank amends the MTA so that it allows sharing the material further.

Respondents were then asked whether such contractual limitations in an MTA are enforceable, and thus practical. Once again, respondents from both the international sample and the US sample substantially confirmed the view that MTAs are indeed practical. Enforcing MTA provisions is as feasible as applying any other contractual obligation, either through informal means or by using domestic legal remedies. Suggestions of possible strategies to monitor compliance were also made. One is practical: to limit the transfer of biological material to other laboratories or outside researchers only to minimal quantities needed for the specified research. Requiring that recipients return all unused biological material was also indicated as a viable way to informally enforce MTA obligations (#57 IS, No. Am./Europe, medicine, university, S: "I want to get everything back, so this is a proof that nothing is travelling, nothing is stored somewhere. I want everything back … I want them to do the analysis to use very parsimoniously the DNA"; #34 IS, Europe, medicine, genetics, biobank, S: "we provide only what is needed [for his/her research]"). During the expert consultation, participants suggested that the control may also be exercised by requiring the authors of publications involving genetic research to state where the samples they used come from. With such a standardized citation method, one would be able to track the use of biobanks or depositories. Moreover, experts pointed out that biobanks also have internal control mechanisms such as the requirement to file an application in writing to access samples, coupled with ethics approval for all applications filed. Finally, peer pressure makes it harder for an investigator with a reputation of breaching MTAs to receive material in the future.

Other respondents indicated formal legal avenues of enforcement are also available to enforce MTA obligations. In particular, legal remedies commonly available to the non-breaching party to a contract were indicated as an effective means of enforcement.

A few respondents found the question beside the point because they held the view that the enforceability of an ethical obligation is irrelevant as to assessing its validity. Ethical principles are valid even when their enforceability is problematic and even if it is unlikely that the ethical obligation will be respected.

3. Requirements imposed on researchers: sharing findings with the biobank

The study also explored the requirement that investigators receiving samples from, or otherwise accessing samples stored in, a biobank share "all research findings and the data produced for each sample with the repository" (scenario 6(b)). Responses to this question must also be read in conjunction with responses to questions that touch on publication and patenting, which were also raised in this study (see chapter 12). As discussed in Chapter 12, respondents supported a requirement that data be placed in the public domain and agreed on the need to scrutinize the boundaries of gene patenting. It follows that the discussion on the requirement to share findings with the biobank from which samples originated is very much informed by these broader views on the duty to publish and not to patent. Indeed, as a logical matter, any requirement that researchers put all research findings and data into the public domain would encompass the obligation to make the data available to the biobank from which samples originated. Moreover, the argument to restrict patenting implies that research data and findings cannot be the subject of private appropriation, but that they must be shared with the other scientists and more generally with the public. However, the two sets of issues are actually distinct, and, as the interviews demonstrate, some respondents find it ethically appropriate to impose a duty to share findings with the biobank in the absence of both a duty to publish and any restriction on patenting rights.

Respondents in both the international and the US samples were in substantial agreement that sharing findings with the repository ought to be required. Several arguments were advanced to support such a policy. First, feeding the findings back to the biobank reduces the risk of duplication of research and therefore it is an efficient, desirable requirement (#35 IS, Europe/So.Am., medicine, genetics, university, S: "If information is shared with and available at the repository, it avoids repeating research on precious samples"). Second, the altruistic motives that ideally underpin participating in scientific research as well as undertaking it offer a justification for this requirement (#94 IS, Asia, life sciences, government, A: "human samples should be considered a common [good] and research findings derived from those samples should be shared ... In principle, those human samples [are] given free of charge for goodwill to research disease genes, I think the ideal goal is the human welfare"). This approach reflects the traditional view that scientific knowledge ought to be shared among scientists. Third, respondents applied the principle of reciprocity between the biobank and scientific investigators: if a researcher uses samples or information collected and stored by a biobank, then he or she ought to give something back to the biobank. While some respondents expressed the view that the requirement that findings be shared should be "taken from granted" (#93 IS, Europe/Middle East, biology, genetics, biobank, S), others argued "it should be more an open arrangement that a contract" (#48 IS, No. Am./Europe, philosophy, university, R/A).

The lack of flexibility in the actual requirement was problematic to some respondents for a number of reasons. First, transferring raw data back to the biobank would increase the likelihood of including also data with material mistakes, data that could lead to no findings, and data that are difficult to organize and utilize at a later time. Second, a rigid requirement would result in a lack of interest on the part of investigators interested in using the findings for commercial purposes. A respondent working in a governmental research institution argued that "[the requirement] seems unrealistic ... one competitor gets all the results generated by [another] competitor" (#17 IS, Europe, law, university R/A). To address these concerns, several respondents indicated that research ought only to have to share aggregate data with the biobanks; further, they might possibly grant the investigators an opportunity to delay sharing their data so that they can analyze the data and publish the findings in scientific journals. Regarding negative findings, the few respondents who addressed this complication mostly concluded that negative findings ought also to be shared because these are perhaps the most interesting findings.

Few respondents expressed the view that no sharing is required. Those who did argue that "the repository is like an infrastructure" (#03 IS, So. Am, nat. sciences, genetics, bioethics, university, A), and consequently a researcher has no obligation to share findings with institutions not engaged in research activities. One respondent added that sharing findings might even result in giving biobanks too much "power."

4. Conclusions

Biobanks are often seen as infrastructures for the research community as opposed to self-contained research programs that aim at studying specific questions. Therefore, biobanks are expected to transfer samples and data to researchers who work at institutions separate from the biobank. As a result of doing research with samples originating from a biobank, these investigators take advantage of the efforts and investments needed to set up the biobank. The respondents' answers reflect a growing consensus that obligations arise out of biobank access. In particular, investigators are required to stipulate the scope of the planned use of the sample and data and are prohibited from transferring the samples to third parties or to use samples and data for uses not agreed upon. Moreover, investigators also have a duty to share researchers' findings with the biobank, so that it is continuously enriched by its use and efficiently exploited. Although the enforcement of these obligations may prove problematic, these ethical duties exist. Legal systems provide ways of enforcing the obligations. Still, a greater role ought to be played by the scientific community in making a responsible use of biobanks.

Bibliography

Merz, J.F., Magnus, D., Cho, M.K. and Caplan, A.L. (2002), "Protecting subjects' interests in genetics research," *American Journal of Human Genetics* 70: 965-71.

Rodriguez, V. (2005), "Material transfer agreements: open science vs. proprietary claims," *Nature Biotechnology* 23, no. 4, 489-91.

Rotimi, C., Leppert, M., Matsuda, I., Zeng, C., Zhang, H., Adebamowo, C., Ajayi, I., Aniagwu, T., Dixon, M., Fukushima, Y., Macer, D., Marshall, P., Nkwodimmah, C., Peiffer, A., Royal, C., Suda, E., Zhao, H., Wang, V.O., McEwen, J. and The International HapMap Consortium (2007), "Community Engagement and Informed Consent in the International HapMap Project," *Community Genetics* 10: 186-98.

PART III
SHAPING THE FUTURE LEGAL
AND ETHICAL DEVELOPMENT
OF GENETIC DATABASES

Chapter 15

Towards an International Framework: Results of a Meeting of an International Group of Scholars and Scientists Involved in Legal and Practical Issues of Biobanks

Alex Mauron

1. Introduction

Early on during the planning of the research project undertaken jointly by the Department of Ethics, Trade, Human Rights and Health Law of the World Health Organization, and the Institute for Biomedical Ethics of the University of Geneva, a meeting of high-level experts in the field was planned, with a view to assessing the results of the project and to place them in the context of current debates. The meeting was convened on May 8 and 9, 2006 at WHO headquarters in Geneva and the following persons attended:

Professor Godfrey Banyuy Tangwa, University of Yaoundé
Dr. Andrea Boggio, Keele University
Ms. Liz Bowie, intern WHO/SDE/ETH
Dr. Anne Cambon-Thomsen, CNRS, Toulouse
Professor Alexander Capron WHO/SDE/ETH
Professor Leonardo De Castro, University of the Philippines
Dr. Bernice Elger, University of Geneva
Ms. Agomoni Ganguli, University of Zurich
Professor Hank T. Greely, Stanford University
Dr. David Gurwitz, Tel-Aviv University
Professor Wayne Hall, University of Queensland, Brisbane
Dr. Robert Hewitt, National University Hospital, Singapore
Professor Bartha Knoppers, University of Montréal
Professor Alexandre Mauron, University of Geneva
Dr. Catherine Moyes, Oxford University
Professor Victor Penchaszadeh, Columbia University, New York
Professor Margarit Sutrop, University of Tartu
Professor Jeong-Ro Yoon, Korea Advanced Institute of Science and Technology, Daejeon.

As the prime purpose was to get feedback on the results of the research project and to debate the issues addressed by it, the sessions were organized around the project scenarios and the main ethical issues they raise. The following summary has the same structure, but also reflects the fact that some issues received more attention than others. The main discussion points are summarized hereafter without attributing comments individually.[1]

2. Summary of the meeting

2.1 General comments

Andrea Boggio and Alex Capron explained the overall methodology of the international study. A purposive sample of 42 international experts in biobanks and the relevant ethics and law were interviewed on the basis of the scenarios presented in chapter 4. The sample of interviewees, albeit coming from a wide range of geographical areas, was not representative and neither was it meant to be. Rather, the study objective was to collect a wide range of expert views from individuals who had pondered these issues at length. Questions about the purpose of the study and its connection with possible WHO guidelines on the subject were clarified. The research was not meant to provide the groundwork for WHO guidelines, its purpose was to review and clarify policy options on biobanks against the background of a great variety of normative texts showing little consistency.

Andrea Boggio specified the aims of the interviews for scenarios A and D. The points being investigated were the following: consent and use of samples, withdrawal, anonymization of samples/data, ownership, requirements/limitations for researchers, feedback to participants and destruction of samples. The results were briefly summarized and can be found in chapters 5 to 12 and 14.

The experts asked about the professional background of respondents, who may have been biased towards biobank research. The answer was that indeed the project was to explore the opinions of a particular professional group who are deeply engaged with these issues, but who also have rather diverse professional experiences. Attempts to recruit patient advocates into the study proved difficult. The sister US study had a slightly higher proportion of bioethicists and individuals involved with policy, as opposed to scientists working with biobanks directly. A general discussion ensued about how to broaden the opinion base in order for future guidelines to be responsive to a wide range of public opinion. It was suggested that the US sample be thought of as a control group for Western ideas, but then professional bioethicists may not always reflect the ideas of biobank participants, a discrepancy that is suggested in the literature (Wendler and Emanuel 2002; Hoeyer, Olofsson et al. 2004). The importance of keeping the informed consent process simple, while building trust in the biobank governance was emphasized. As regards the international study, it was also noted that a polling approach is less valuable than a more in-depth inquiry that elicits responses from individuals who are genuinely concerned with the issues.

---

1 This summary is based on notes taken by Liz Bowie.

It was noted that some responses in the study may have been a reflection of the legal framework in the respondent's country. This generated a general discussion about the role of legislation in this field, which revealed positive expectations towards the law, for instance that outlawing genetic discrimination may defuse criticism or fear of population genetics. On the other hand, ethical concerns cannot be wholly subsumed by law even though the legal expert will bring in specific questions, such as the consistency of rules. But the broader questions remain, of how the law reflects prevailing concerns of citizens, how it responds to changes in this fast-paced research area, and how it manages the reconciliation of individual rights and group claims, an issue that is particularly important and contentious in the biobank field.

2.2 Sample coding, withdrawal, ownership

The issue of withdrawal subsumes many of the broader issues relating to informed consent. As such, it received a great deal of attention, and the difficulty of applying the clinical trial model to biobank research was emphasized. On one view, participation in biobank research is really a donation process and withdrawal of samples should not be possible. In addition, there is a conflict between the right to withdraw as construed by traditional research ethics since the Nuremberg Code and the Helsinki Declaration, and an alternative notion according to which the biobank research participants enter a kind of contract from which they cannot withdraw at will. The classical view of research ethics was tailored to the specifics of clinical trials and did not anticipate the burden that withdrawal places on biobank research. Furthermore, to construe withdrawal as an appropriate remedy for privacy violations seems questionable. Privacy should be protected by specific measures, rather than upholding the right to withdrawal as a kind of "punishment" for breaches of privacy. On the other hand, it would seem that if a sample donor is dissatisfied with the information received or with the research done with her sample at a later time, the right to withdraw is an ethical necessity. The lawsuit brought by the Havasupai tribe against Arizona State University (Dalton 2004), as well as the case of William Catalona v. Washington University, St. Louis (Andrews 2006), were brought into the discussion and show that withdrawal becomes a pressing issue in case of severe conflict (in the Catalona case, samples anonymization was used a as "remedy" when participants wanted to withdraw). At any rate, it is essential that biobank research participants are made aware of the withdrawal options that are open to them and that these should also be factually possible. It was also noted that the possibility of withdrawal may be a prudential matter, since without it participant recruitment may be more difficult. Also, sample donors may have various reasons to want to stop their participation in a biobank, reasons that need not be connected with deceit or misconduct on the part of the researchers, but with the possibility that a new, initially unanticipated research project meets with disfavor from the sample donor. In this respect, the right to withdraw seems a reasonable protection against an uncertain future.

The practical consequences of withdrawal were also discussed and raise the question of whether samples (and possibly also unpublished data) should be destroyed or anonymized. There seems to be a consensus among experts that irreversible anonymization is detrimental to scientific research and that its ethical benefits are questionable, both because the public often misunderstands the distinction between coding and anonymization, and because absolute anonymity is to a large extent an illusion. In practice, double coding is preferable. It makes longitudinal studies possible while affording a similar privacy protection as anonymization since it prevents researchers from identifying samples on their own.

Another way of managing new research objectives while mitigating the need for a right to withdraw is ongoing review by an IRB or ethics committee. This led to a controversial discussion reflecting the rather different public perception of ethics committees in different countries. While the IRB system seems relatively well accepted in some countries, including the US, ethics committees are often felt to be too researcher-friendly. Alternative proposals involve patient organizations, or other neutral non-governmental bodies. Again, as in other sessions of the meeting, it was apparent that what counts as good governance will vary from country to country.

Finally the discussion of ownership highlighted the contrast between the substantial support among study respondents for sample ownership by research participants, and expert opinion, which favors ownership of samples and data by the biobank. The latter opinion is now dominant and thought to be more practical as well as better suited to the prevention of abuses. The subtleties of the legal property concept may be lost on non-lawyers, and this may complicate the discussion of property rights and custodian responsibilities of biobanks.

2.3 Dissemination of data and intellectual property (IP) rights

The study respondents agreed that researchers using data and/or samples stored in a biobank must inform the biobank of their findings. This view was also shared by the experts, who emphasized the principle of reciprocity that must operate between the biobank and researchers. In addition, if helping the population is listed as a benefit of biobank research, this entails a commitment to publishing research results. The public benefits if biobanks centralize research results that originate from their samples, because this may help disseminate new findings and avoid duplication. However this assumes that biobanks take on what is in effect an additional distinct role, namely to build public databases. It is not obvious that biobanks, as sample repositories, are necessarily the best equipped to construct and maintain such public databases. On the other hand, it is clear that a biobank is more than a mere storage facility and that its management involves an active engagement with research. Another related practical issue was mentioned, namely that there is an increasing tendency in biobank research to "economize" samples, in other words to manage sample collections with a view to rescuing as much information from them as possible, rather than gathering more samples. This feeds back into the discussions of informed consent, because it puts a premium on obtaining an open-ended consent for further research.

The current consensus in favor of data sharing was reinforced by the historical precedent of the Human Genome project and the Bermuda agreement, which mandated virtually instant publication of sequence data (Marshall 2001). The experts noted that it is not clear that this high standard of public sharing reflects a general agreement in the scientific community when it comes to current biobank research.

Some experts agreed with respondents in the study who believed that researchers have a right to delay the release to the public in order to secure publication in a scientific journal and possibly IP rights. It appears that the different responses reflect the broader controversy of IP rights in health-related research. For some, IP rights in biomedical research have gone too far, and they argue that the IP system creates impediments for further research and also puts developing countries at a disadvantage (an issue that should be of major concern for WHO). For others, a solid system of IP rights is *sine qua non* to persuade private companies to undertake research, and if biobanks require researchers to abstain from patenting, then companies will not collaborate. It was noted that IP issues are complicated by the fact that the relevant laws differ considerably from country to country. In addition, important distinctions in IP legal doctrine, such as the difference between use patents and composition-of-matter patents, seem often lost in these controversies.

The study addressed the issue of Material Transfer Agreements and this led the experts to discuss the duties of researchers as regards biobank samples and data. Many biobanks have rather stringent regulations as to the use of their samples. Upon application, researchers must describe the purpose of research and the process of ethical approval. Therefore biobanks exert some degree of control and are in a position to assess whether researchers have the appropriate infrastructure and funding and whether they are able to adhere to privacy protection measures. Nevertheless that control seems imperfect and according to one expert, it is nearly impossible to track the use of samples at the moment. This raises a closely related issue, namely the problem of acknowledging biobanks in research reports. There should be more standardization in this respect, along the lines of the rules established by scientific journals for citation and acknowledgement.

2.4 Feedback to participants

Among possible forms of feedback to participants, there is a need to distinguish scientific results and individualized information of diagnostic or predictive relevance for a given individual. This distinction between knowledge of general importance and personally relevant information is often not made with sufficient clarity.

In biobank research as in biomedical research generally, one usually assumes altruism rather than personal gain as the driving motive of research participants. In exchange for their participation, it is ethically appropriate for subjects to receive information about the results. This much seems obvious, but is often overlooked in practice. To perform research and then fail to let results filter back to participants is to treat participants with levity. Worse still, data are often collected in the developing world to be analyzed in developed countries, which diminishes the incentive to give feedback to the research participants and their communities. There should be more insistence to have data analyzed where they are collected, a practice that would foster research capacity building in developing countries.

Experts were more divided about individualized feedback to participants, which raises a host of practical, ethical and legal issues. On the face of it, to provide individual participants with relevant health-related information discovered in the course of research seems reasonable enough. It reflects a dual "right to know and not to know" that is well established and originated with the ethics of genetic counselling. However, this raised a discussion of who is best qualified to provide this information to participants and to cope with the consequences in terms of providing counselling and treatment. As regards competence, some experts expressed doubts that general practitioners are sufficiently qualified to provide genetic counselling. The case of the Estonian database was mentioned, which initially foresaw that all participants will have their personal data explained to them by their physician. But in the absence of resources earmarked for this service, false expectations were raised in the public. It also seems disingenuous to persuade the public that it will benefit from pharmacogenomics, only to discover that, if a treatable condition is diagnosed, treatment is unaffordable or not provided by the health care system. More generally, a general requirement that biobank research must entail the ability to recontact individual participants with personal information may be counterproductive, if it makes some valuable research too costly or downright impossible (the case of research on tropical diseases in poor countries was mentioned).

What personal feedback options appear appropriate seems to depend rather strongly on the cultural and legal context. In the US, scientists and administrators are sceptical in this respect, because they envision the various legal liabilities that an obligation to inform might entail. Furthermore, there is a separate regulatory framework for clinical laboratories as opposed to research laboratories. Only those who are certified under the Clinical Lab Improvement Act (CLIA) can return results to individuals, although research labs may seek CLIA certification (a situation that may exist in other developed countries). Two Swiss biobanks were mentioned who foresee that incidental findings will be communicated to the participant's treating physician, who will then initiate contact with the individual. This was perceived to be largely unthinkable in the US context. Several experts suggested that a measure of discretion should be left as to the extent and manner of personal feedback. Also, there seems agreement that whatever the personal feedback possibilities in a given project, these should be explained clearly to research participants without raising unrealistic hopes. In addition, if one anticipates that in certain cases (an incidental finding with life-and-death consequences, for instance) the terms of the initial consent may be overridden, that fact should be mentioned to participants.

2.5 Destruction of samples, terminating a biobank

The long-term fate of biobanks raises issues of its own. There are precedents for sample collections to be sold to private companies, destroyed, or transferred from private ownership to a public institution for preservation. This ties in closely with the issues raised by open-ended consent. In the US study, some respondents noted that they assumed that if they should consent for further research, that would mean that their samples be used for a long time, possibly well past their own demise. Nevertheless, as experts noted, truly indefinite storage may not be a realistic proposition and may run into legal obstacles. Still, long-term storage and the ability to do research long after sample donation is an important asset of biobank research. Given the current trend to use samples sparingly, as mentioned earlier, it appears essential to have informed consent practices consistent with such long-term research use, as well as opportunities for supplementing the initially recovered health data by new information obtained from the sample source's treating physician or other health care providers.

2.6 Territorial limitations, "common heritage," benefit sharing

Agomoni Ganguli presented the results of the study pertaining to scenarios B and C (see chapters 7, 13 and 14). She cautioned experts against excessively literal interpretations, given the language barriers and the possibility that some responses are biased by "social desirability."

The discussion focused largely on the tension between territorial restrictions to protect a country's stake in genetic data and the common heritage approach to the human genome. The latter is not uncontroversial, since it originated with natural resources deemed to be a "common heritage" in order to protect them from overexploitation. It is not obvious that this language is readily applicable to the human genome. A related, but distinct debate concerns genetic resources materialized in plant and animal genetic diversity, where the economic language of resource exploitation and sharing makes more sense, and where developing countries have been vocal in fighting to retain IP rights, for instance to protect the value of local knowledge. Also, territorial restrictions on floristic and faunistic resources are meant to foster capacity building in developing countries, thereby making room for more egalitarian scientific collaborations between developed and developing countries. It was mentioned that such laws, passed for instance in China, are meant as temporary stopgap measures against bio-piracy.

Bio-piracy is a relevant concept when one takes plants and useful plant-lore away from developing countries to develop commercially viable applications without sharing benefits. In contrast, taking human DNA samples away without addressing the human and natural environment of local populations makes little sense. Therefore to work with developing world populations and to share capacity with developing world scientists is not only ethical, but also good scientific practice. In addition, when researchers in the Western world turn to developing countries and their resources, this need not be malevolent. The example of HapMap was mentioned, where it would in fact have been simpler and cheaper to gather samples in the US only, but where the decision was made to go to developing countries in order to provide benefit-sharing and capacity building.

Capacity building has another indirect benefit. Experiences in India for instance show that some form of territorial regulation (rather than prohibition) on research is useful, if it fosters the participation of local scientists. In addition, WHO is developing an instrument to assess a country's health needs, so that research can be responsive to the health needs of the community. Researchers from the Western world would then have to justify why they do their research in a developing country. Nevertheless, experts highlighted the need for additional efforts in this respect. It was noted that Western labs hire PhD students to collect material in developing countries and then consider these to be their own. This leads to considering the importance of a kind of benefit-sharing distinct from commercial gain, namely the benefit resulting from scientific advances relevant to local health problems. Communities who provide consenting subjects to research, and maybe take personal risks in participating, should also reap the benefits, and these should also be understood in terms of this knowledge-based utility, as it were, not just money.

Finally, the ethical dilemmas raised by a particular form of benefit sharing common in the developing world were discussed, namely the provision, linked to the establishment of a research project, of general medical services not normally available to the study population. Problems are obvious if that health care effort is not sustainable in the long run. But even if this aspect is addressed satisfactorily, the very nature of such a benefit-sharing scheme may be seen by some as a subtle inducement that may compromise the validity of consent.

Bibliography

Andrews, L. (2006), "Who owns your body? A patient's perspective on Washington University v. Catalona," *J Law Med Ethics* 34:2, 398-407.

Dalton, R. (2004), "When two tribes go to war," *Nature* 430, 500-2.

Hoeyer, K. and Olofsson, B.O. et al. (2004), "Informed consent and biobanks: a population-based study of attitudes towards tissue donation for genetic research," *Scand J Public Health* 32:3, 224-9.

Marshall, E. (2001), "Bermuda rules: community spirit, with teeth," *Science* 291, 1192.

Wendler, D. and E. Emanuel (2002), "The debate over research on stored biological samples: what do sources think?," *Arch Intern Med* 162:13, 1457-62.

Chapter 16

Biobanks and Genomic Research: What Shape the Future?

Alex Capron

1. Bringing the parts together

This book brings together three lines of investigation. One takes stock of the ethical controversies that have surrounded research with human subjects, particularly research involving genetic and genomic studies. The second reviews the existing guidelines, laws and regulations that have been adopted in the past fifteen years to govern genetic databases (or biobanks) and the research conducted using the human biological materials and related data these scientific resources contain. The third reports the results of more than eighty interviews with experts from around the globe and analyzes the meaning and pattern in these responses. The first two parts manifest a great deal of confusion, controversy and contradiction. It would be nice to be able to report that the third replaces confusion with clarity, controversy with consensus, and contradiction with consistency, but at most, the respondents moved us a little way towards those fine objectives, but in the process also added nuance to the generalities of declared principles and demonstrated the practical problems that can confound even clear and simple rules.

Yet, like a stool with three legs, there is a way that the three parts of this study come together to form a whole. Of course, from one viewpoint, this unity is hardly surprising since the "scenarios" used to probe the respondents' views were largely built around the issues raised in the ethical and regulatory analyses. Yet from another viewpoint, one might have predicted that the geographical and professional diversity of the respondents would lead their views to spiral out in many diverse directions. Instead, at least on some points, the degree of consensus was larger than one might have anticipated and was not fixed either by professional background or geographical location. Whatever one might expect, this turns out not to be a story of South versus North, nor of science versus society. For instance, while the terminology continues to cause problems, irreversibly anonymizing stored samples struck most respondents as a poor idea. Likewise, respondents were generally favorable to explicit plans for benefit-sharing rather than payments to individual participants for their samples; relatedly, they thought that researchers were at work on an enterprise that should benefit the community and hence have an obligation to return their results to the biobanks as a means of placing them in the public domain.

Second, where confusion still exists, the need for clarification seems generally accepted. The respondents admit some confusion about the concepts of coding and anonymization of samples, and about different attributes of property, both tangible and intellectual. Resolving points of confusion would probably produce greater substantive consensus.

Third, even people with different substantive views seemed implicitly committed to fair and transparent processes to resolve such differences; rather than hard, ideological stances, the respondents displayed an awareness that the issues in question are ethically controverted and not easily resolved. For example, the wide agreement on the role that Material Transfer Agreements can play in exerting control over the use of genetic samples illustrates a procedural approach, even if it leaves open the question of whether it is normatively desirable to limit the circulation of DNA or tissues by national boundaries or to researchers working only on certain topics.

Finally, the issues that gave the respondents the most trouble—how to balance individual and community interests, whether commercialism is appropriate in genomics research, and what role traditional ethical obligations to research subjects should play in the setting of biobanks—are the very ones most poorly addressed by existing guidelines. This suggests that if the people responsible for such documents— both in the form of ethical declarations and in the form of regulations—pay attention to the fine distinctions and detailed issues perceived by people "in the field" (whether researchers, biobankers, bioethicists, lawyers, community advocates, or others) the documents that they draft may be able to broaden the base of agreement across these groups.

2. A gradual normalization

At the heart of the confusion, controversy, and contradictions, and of the potential clarity, consensus, and consistency is the odd role that genetics generally and genomic mapping in particular play in modern societies. As Alex Mauron noted in chapter 1, fifteen years ago, the deciphering of the human genome was, on the one hand, the "Holy Grail" of biology (Gilbert 1992) and, on the other, a great challenge to accepted ethical norms (Chadwick and Berg 2001). Ironically, both of these visions bought into a questionable, deterministic view of the human genome. The human genetic code was only a sacred thing if it was the "blueprint" of human beings, which of course it wasn't, and the genome mappers soon learned the error of describing the genome in terms that suggested that once it was known, one knew the person "encoded" in the genome. If a metaphor was needed, a musical score is more apt: the notes on the paper are important in determining the music we hear when musicians perform, but the particular skills of the players and, most important, the choices made in particular circumstances by the conductor, are enormously influential in determining what we the listeners will hear at the concert. Likewise, from the vantage point of the ethicist, all the concerns about the human genome were only justified if knowing the genome really meant knowing a lot about the person and his or her future. Both of these views then depended fundamentally on genetic determinism. Yet part—a very large part—of the reason for engaging in genomic research, using genetic databases or biobanks, is to understand more clearly the role that the environment and personal lifestyle play in health and disease, in other words, the degree to which genes do not determine—in the sense of setting in stone—who we are.

Today, while genetic determinism is still a cultural force to be reckoned with, it seems much less widely or unthinkingly accepted than it was fifteen or twenty years ago. The forces feeding determinism are often "science-based," at least in the sense that they emanate from the publicity departments of universities happy to proclaim that their physicians have "found the gene for X" (rather than admitting that they don't fully understand the ways in which the identified gene interacts with the environment, which is why everyone with the gene should not immediately assume they'll have the disease). Yet despite such influences, determinism doesn't hold the sway it once did. This can be attributed to greater familiarity with genetics in general culture; people accept that their physician in working up an illness will want to know something about a person's family history, without assuming a one-to-one correlation between the disease one's parent or grandparent suffered from and the symptoms one has today. Likewise, the greater familiarity of tests either for genes or gene-products in ordinary medicine have moved this field out of the speciality clinics (where a few medical geneticists counseled families which had experienced unusual metabolic and comparable disorders) into ordinary clinical practice, making "genetics" less frightening in the process.

Reading between the lines of the responses to the scenarios in this study, I detect that the normalization of genomic research arises also from recognizing that the construction of large biobanks is the creation of a public good of inestimable value. Thus, one of the central, unresolved questions—as can be seen in this book—is what role commercial interests, such as pharmaceutical companies, should play in the development of genomic knowledge. Respondents generally favored putting material in the public domain, but they were not adverse, in an absolute sense, to allowing intellectual property claims to be asserted, at least in ways that did not block group or communal interests regarding the material.

Even speaking of human genetic material is a reminder of the hot debates that have taken place over whether to regard genetic information as material, in the sense of property, or as something more immaterial, such as a person (Chadwick 2001). Yet today, genetic material seems to be coming to be viewed, as Mauron remarked earlier in this volume, as "stuff." What was once foreign and frightening comes to be seen over time as routine and nonthreatening. In the case of biobanks, the process of normalization seems to be helped by the widespread campaigns to recruit participants. Genomics can no longer attract attention on an "exceptionalism" basis when genomic research is a common part of many epidemiological studies. Furthermore, the association of such research with public objectives—specially public health activities, such as surveillance—only serves to further normalize it.

Of course, normalization does not necessarily mean a disappearance of all concerns. In an era when people are increasingly worried about the misuse of information about themselves on account of greater access to private records through electronic media, normalization could mean nothing more than treating genomic studies the same as any other medical data—that is, making available private information through sloppiness or leaks or lack of well-defined status. But even so, in this scenario, genetic data are not exceptional.

3. International perspectives

At the heart of this study is a question about the role of universal ethical duties and rights (tied even to human rights) versus national rules, which may be embodied in domestic legislation and regulations. The drive toward universal standards is very strong, not the least because the underlying scientific activity itself cuts across continents. The assumption of modern scientists is that common standards should prevail around the world. To the extent that national standards must fit within domestic legislation, it is easy to expect that a conflict will arise, but in the end, international norms can affect the shape of domestic regulation in a fundamental fashion.

Among the topics that seems most to divide the respondents, and probably the population at large, is the role of commercialism in genomic research. The history of exploitation of natural resources generally—including recent pharmaceuticals developed using traditional knowledge (with and without the agreement of, and the sharing of benefits with, the communities that had cultivated the plants or animals in question for generations, and that had demonstrated their therapeutic value) — make some people very wary of anything that seems to resemble "gene prospecting" among isolated populations, whatever good genetic reasons may be cited for the practice. Further, the broader debates about patents on vital medicines held by biotech and pharmaceutical companies, and the adverse effects these patents have on access by poor populations to essential medicines and vaccines, also complicate any discussion of commercialization of genomic databases. On the other hand, the first major national "biobank" (in Iceland) was both an expression of national pride and unity and an entrepreneurial project. As population databases, such as those in Estonia and the United Kingdom, come to be generally accepted, the "public good" character of biobanks may come more strongly to the fore. If that happens, and is mirrored by a general acceptance among researchers of an "open source" approach to the products of genetic databases, it is likely that the public will come to be less concerned about genetic databases, assuming adequate protective legislation has been adopted. In the end, neither model—public or commercial—necessarily rests on either the "property" or the "person" approach to genetic material; instead, what people seem to be most concerned about is the extent to which their legitimate expectations that their interests will be protected and that, to the extent possible, they will be able to participate in controlling what is done with information about themselves (Rothstein 1997). The abstractions about "owning" people are likely to fade, not the least because they are not really needed to resolve the question of control. In a post genetic-determinism world, a human biological sample is just that, not the keys to the kingdom of every person's essence much less his or her future.

Charting the future will require developing much more "local knowledge" of the sort that the study described in this volume began to develop. A larger and even more thoroughly international study could help to flesh out the issues that actually confront people operating, or conducting research with, or thinking of donating to biobanks. As we have seen, the existing international guidelines and declarations do not answer anywhere near all the important issues that arise for these players on-the-ground. On the other hand, these individuals are also not eager to be cast off on their own devices, but rather see value in global deliberations and the elaboration of guidance that is responsive to the problems that they confront in their work. The early work has been done, and now is the time—with a more realistic view of the science and of its social and moral import—to develop ethical and legal documents that are more fully attuned to the perspectives of people around the globe.

Bibliography

Chadwick, R. (2001), *Informed Consent and Genetic Research* (London: British Medical Journal Books).

Chadwick, R. and Berg, K. (2001), "Solidarity and equity: new ethical frameworks for genetic databases," *Nat Rev Genet* 2:4, 318-21.

Gilbert, W. (1992), "A Vision of the Grail," in Kevles, D.J. and Hood, L.E., *The Code of Codes: Scientific and Social Issues in the Human Genome Project* (Cambridge, Mass.: Harvard University Press).

Rothstein, M. (1997), *Genetic Secret* (New Haven: Yale University Press).

Index